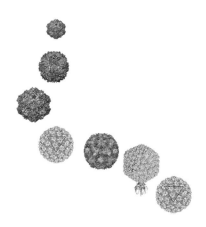

VIRAL
NANOPARTICLES

TOOLS FOR MATERIALS SCIENCE & BIOMEDICINE

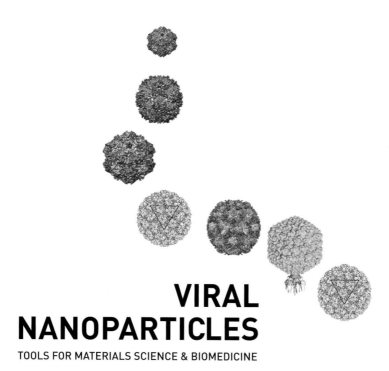

VIRAL
NANOPARTICLES

TOOLS FOR MATERIALS SCIENCE & BIOMEDICINE

NICOLE F STEINMETZ
MARIANNE MANCHESTER

PAN STANFORD PUBLISHING

Published by

Pan Stanford Publishing Pte. Ltd.
Penthouse Level, Suntec Tower 3
8 Temasek Boulevard
Singapore 038988

Email: editorial@panstanford.com
Web: www.panstanford.com

British Library Cataloguing-in-Publication Data
A catalogue record for this book is available from the British Library.

VIRAL NANOPARTICLES: TOOLS FOR MATERIALS SCIENCE AND BIOMEDICINE

ISBN 978-981-4267-45-8 (Hardcover)
ISBN 978-981-4267-94-6 (eBook)

Printed in Singapore by Mainland Press Pte Ltd.

Contents

Preface

Viral nanotechnology is a young and emerging discipline. A highly interdisciplinary field, viral nanotechnology inhabits the interface between virology, chemistry, and materials science. The field exploits viral nanoparticles (VNPs) for potential applications in diverse fields that range from electronics and energy to novel materials and medicine.

From a materials science point of view, VNPs are attractive building blocks for several reasons. In general, VNPs are on the nanometer-size scale; they are monodisperse with high degree of symmetry and polyvalency; they can be produced with ease on large scale; they are exceptionally stable and robust; and they are biocompatible, and in some cases, orally bioavailable. VNPs are "programmable" units that can be modified by either genetic modification or chemical bioconjugation methods.

This book will give a survey of the applications of VNPs in fields ranging from materials science to biomedicine. An introduction to the field of viral nanotechnology is given in Chapter 1. Chapter 2 provides an overview of the many different VNP building blocks currently in use for viral nanotechnology. Various methods have been established that allow efficient production of VNPs as well as their non-infectious counterparts, the virus-like particles (VLPs), as discussed in Chapter 3. A large variety of bioconjugation methods have been applied and optimized for VNPs that facilitate chemical modification, fine-tuning, and immobilization of VNPs; these techniques are described in Chapter 4. Chapter 5 summarizes strategies to entrap or encapsulate non-natural cargos into VNPs and VLPs. Mineralization chemistries are discussed in Chapter 6. With the different chemistries in hand, a range of highly interesting VNP-based materials has been fabricated with potential applications in sensors, electronics, and medicine. Toward the development of nanoelectronic devices, Chapter 7 describes the efforts that have been directed to the generation of thin-film arrays of VNPs immobilized on solid supports. Last but not least, Chapter 8 provides an overview of the advances of designing VNPs for biomedical applications, including their uses as vaccines, imaging modalities, targeted therapeutics, and gene therapies.

As this young discipline matures, a new era has begun in which pathogens have become useful material building blocks for next-generation nanodevices. The field has even greater promise given the enormous variety of available VNPs with their vast diversity of sizes, structures, chemical reactivities, and biocompatibilities. This promise drives an inspiring field with wide-ranging opportunities for discovery.

La Jolla, CA, December 2009   *Nicole F. Steinmetz, PhD*

Marianne Manchester, PhD

Chapter 1

AN INTRODUCTION TO VNPs AND NANOTECHNOLOGY

Viruses have long been studied as pathogens, with the goal of understanding viral infection and disease. More recently viruses have begun to be regarded as building blocks and tools for nanotechnology. Viruses are exploited as *platforms*; that is, they are used as templates or scaffolds for the design of novel nanomaterials. A wide variety of viral platforms have been studied and utilized for applications ranging from materials to medicine. This chapter will provide an introduction to the role of viral nanoparticles (VNPs) in nanotechnology.

1.1 WHAT IS NANO?

Nanotechnology is a highly interdisciplinary field that brings together researchers from different scientific backgrounds and has created a novel common language. It is a collective term for a broad range of novel topics concerned with matter on the nanometer scale. Nanotechnology sits at the interface of biology, chemistry, physics, material science, and medicine. Nanotechnology is found and applied in nearly every scientific area.

Nano is a somewhat fashionable term; in common speech, it is used as a prefix to denote something that is smaller than usual. As of this writing, the term "*Apple iPod nano*" is the first hit on a *Google* search of the term "nano." The word nano is derived from the ancient Greek word for dwarf. In a scientific context the prefix nano is used to describe "a billionth of something." A nanometer is a billionth of a meter (10^{-9} m = 1 nm), and a nanosecond is a billionth of a second (10^{-9} s = 1 ns).

Figure 1.1 illustrates how tiny nano is. An aphid insect (Fig. 1.1, panel A) is about 1 mm in size. The aphid is about 1,000,000 times bigger than a nanometer. Human hairs (Fig. 1.1, panel B) have an average diameter

Viral Nanoparticles: Tools for Materials Science and Biomedicine
By Nicole F. Steinmetz and Marianne Manchester
Copyright © 2011 by Pan Stanford Publishing Pte. Ltd.
www.panstanford.com

of 100 µm and are thus 10 times smaller than the aphid and still 100,000 times bigger than a nanoparticle. Panel C shows a plant cell, which is about 10 µm in size and thus 10,000 bigger than a nanometer. Particles formed by the plant virus *Cowpea mosaic virus* (CPMV; Fig. 1.1, panel D) are about 30 nm in diameter and thus nanoparticles. One would need around 50,000,000 VNPs to fill up the interior of a cell.

A single atom is a fraction of a nanometer in size; molecules, including biological molecules, are typically nanometers in size and can thus be regarded as nanoobjects. In recent years a range of biological molecules have been exploited for nanosciences and nanotechnology. Nucleic acids, for example, are used as construction materials to generate highly ordered 2D and 3D structures and assemblies such as nanotubes and nanocages. A main theme in nanotechnology is controlled self-assembly, with the goal being to generate functional materials with a high degree of precision. Nanotechnology, then, requires chemical and physical control at the molecular level. Nucleic acids, proteins, and viruses are essentially naturally occurring nanomaterials capable of self-assembly with a high degree of precision. This property, coupled to the relative ease of experimentally controlling and producing biological nanomaterials, has led to tremendous interest in their nanotechnology applications. Viruses, and VNPs in particular, possess a number of traits that make them exceptionally outstanding candidates.

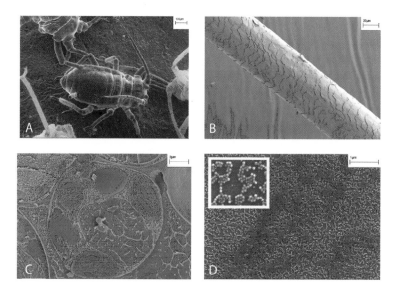

Figure 1.1 Scanning electron micrographs of aphids on leaf (A), human hair (B), fractured plant cell (C), and *Cowpea mosaic virus* particles (D). Panel A–C provided by courtesy of Kim C. Findlay, John Innes Centre, Norwich (UK). Panel D from Steinmetz, N. F., et al., unpublished.

1.2 WHERE DID IT ALL BEGIN? A HISTORY OF VNPs: FROM PATHOGENS TO BUILDING BLOCKS

The word *virus* is Latin and means "poison." Viruses are infectious agents, and generally pathogens. It was not, however, until the end of the 19th century that viruses were discovered as infectious agents. The first virus to be recognized as an infectious agent distinct from bacteria was the plant pathogen *Tobacco mosaic virus* (TMV) (Zaitlin, 1898). Today more than 5,000 viruses have been discovered and described, although this likely represents a fraction of those found in nature. Viruses cause many human diseases, from the common cold and chicken pox to more serious infections such as AIDS (acquired immune deficiency syndrome, which is caused by the *Human immunodeficiency virus* [HIV]) and SARS (severe acute respiratory syndrome, which is caused by SARS coronavirus). Virology — the science of studying viruses — is thus a highly important discipline in regard to human health.

Viruses infect all forms of life. Generally, animal viruses infect animals, including humans; plant viruses infect plants; and bacteriophages infect bacteria. Archaeal viruses are those that infect Archaea. Archaea show similarities with bacteria as well as with eukaryotes, and although they are prokaryotes, it has been suggested that they are more closely related to the eukaryotes (Woese & Fox, 1977).

In their simplest form, viral particles consist of a nucleic acid genome and a protective protein coat termed the capsid. Some viruses have additional structural features such as a lipid envelope, or they may consist of separate head and tail structures (discussed in Chapter 2). In brief, the nucleic acid genome encodes the genetic information that is needed to produce viral progeny. In addition to cellular attachment, an important function of the capsid of non-enveloped viruses is the protection of the nucleic acid genome. This tends to make non-enveloped viral particles extremely robust. With a few exceptions, nearly all viruses utilized in nanotechnology are non-enveloped particles. The envelope for enveloped viruses also plays a role in the initial stages of the infection process, including binding to surface receptors and internalization into the host cell.

Viruses have now been studied for more than 100 years, and detailed knowledge about the structure and function of many viruses has been gathered. For many years the emphasis has been on the understanding of viral infection and disease, and it still is. Being able to control or treat viral infections is an important goal in human medicine (as well as veterinary medicine and agriculture). Every year novel viruses or virus strains evolve with the potential to cause disease and death worldwide. For instance, at the time of writing this book, the *Influenza* virus strain H1N1 (also referred to as "swine flu") has emerged and quickly spread all over the world. The science of fundamental virology will always play an important role in medicine.

By the 1950s, researchers had begun thinking of viruses as tools in addition to pathogens. Bacteriophages, for example, played a key role in the development of molecular biotechnology. Bacteriophage genomes and components of the protein expression machinery have been widely utilized as tools for understanding fundamental cellular processes such as nucleic acid replication, transcription, and translation. Virus genomes are small and the genetic elements that control expression of the genome are highly efficient and multifunctional. On the basis of these properties, several viruses have been exploited as expression systems in biotechnology. Several cloning vectors are derivatives of bacteriophages, and typical examples include the *Escherichia coli* phages λ and M13. Various phage-encoded promoters (DNA sequences that facilitate transcription of DNA into RNA) have been utilized to regulate gene expression. An overview of tools used for molecular biology can be found in *Molecular Cloning: A Laboratory Manual*, by Sambrook and Russell (2001). The use of viruses as cloning and expression vectors is not restricted to phages; plant viruses, insect viruses, and mammalian systems have also been engineered for these purposes. Some of these systems are discussed in Chapter 3.

Another early application evaluated was *bacteriophage therapy*, the use of bacteriophages to combat bacterial infections. With the development of antibiotics (compounds that kill bacteria), which have proven to be more efficient and comprehensive compared with bacteriophage therapy, few efforts were made toward its further development. Other applications include *bacteriophage-mediated microbial control* (applied in the food industry) and *phage display technologies* that allow screening for biological protein-binding partners. More recent developments include the use of bacteriophages for *vaccine* production and *gene delivery* approaches. Developments in using bacteriophages for biotechnological applications were reviewed by Clark and March (2006) and Marks and Sharp (2000).

In the 1970s many efforts focused on the production of virus-like particles (VLPs) for use in anti-viral vaccines (reviewed in Garcea & Gissmann, 2004; Grgacic & Anderson, 2006; Ludwig & Wagner, 2007). A VLP is a particle consisting of the capsid but lacking the genome. A VLP is the replication-deficient and thus non-infectious counterpart of a VNP. Chimeric VLPs and VNPs have also been designed. A chimera is a genetically modified version of a naturally occurring particle or cell. In vaccine development, chimeras are used as carriers or platforms for the presentation of antigenic sequences (sequences that induce an immune response) of other pathogens (reviewed in Garcea & Gissmann, 2004; Grgacic & Anderson, 2006; Ludwig & Wagner,

2007). More details and insights on the use of viruses in vaccine development are given in Chapter 8.

In the 1980s researchers began exploiting plant viruses as expression vectors (a DNA-based plasmid that promotes the expression of foreign genes) to produce pharmaceutical proteins in plants. Advantages of protein production in plants are the absence of contamination with animal products, low production costs, and — when using viral expression vectors — achievement of high expression levels. A range of pharmaceutically relevant proteins including therapeutic antibodies have been successfully produced using viral vectors such as TMV, CPMV, and *Potato virus X* (PVX) (Awram *et al.*, 2002; Canizares *et al.*, 2005; Johnson *et al.*, 1997; Porta & Lomonossoff, 1998; Scholthof *et al.*, 1996).

Viruses Became VNPs. Beginning about 20 years ago, the focus on exploiting viruses and their capsids for biotechnology began to shift toward using them for nanotechnology applications. Douglas and Young (Montana State University, Bozeman, MT, USA) were the first to consider the utility of a virus capsid as a nanomaterial (Douglas & Young, 1998). The virus of interest in their studies was the plant virus *Cowpea chlorotic mottle virus* (CCMV). CCMV is a highly dynamic platform with pH- and metal ion-dependent structural transitions (see Section 2.2.2). Douglas and Young made use of these capsid dynamics and exchanged the natural cargo (nucleic acid) with a synthetic material, in this case encapsulating the organic polymer polyanetholesulfonic acid. Since then many materials have been encapsulated into CCMV and other VNPs (discussed in detail in Chapter 5). The system was further engineered to allow not only the entrapment of materials but also the size-constrained and spatially controlled synthesis of materials within both the capsid and other protein cages (discussed in Chapters 5 and 6). A protein cage is a hollow, generally spherical protein structure that is typically assembled by multiple copies of protein monomers and thus has similarities to a viral capsid.

At about the same time, the research team led by Mann (University of Bristol, UK) pioneered a new area using the rod-shaped particles of TMV. The particles were used as templates for the fabrication of a range of metallized nanotube structures using mineralization techniques (Shenton *et al.*, 1999). These techniques have received great attention during recent years. In particular the contributions of Belcher and colleagues at the Massachusetts Institute of Technology (MIT, Cambridge, MA, USA) led to the development of a new technology that allowed for the generation of a large range of mineralized nanotubes and nanowires for use in batteries and data storage devices (Lee *et al.*, 2009; Nam *et al.*, 2006; Nam *et al.*, 2008). Mineralization and metal deposition techniques as well as their potential applications are discussed in Chapters 6 and 7.

A third direction began a few years later. In 2002, the first study was reported in which bioconjugation chemistries had been applied to a VNP. The research teams led by Johnson and Finn (The Scripps Research Institute, La Jolla, CA, USA) showed, in a proof-of-concept study, that small chemical modifiers such as organic dyes and nanogold particles could be covalently attached to the surface of CPMV. Attachment and display was achieved with atomic precision (Wang *et al.*, 2002). Since then, various chemistries ranging from standard techniques utilizing commercially available reagents to complex and advanced reactions have been developed (discussed in Chapter 4). The establishment of a wide variety of bioconjugation protocols for VNPs was an important development and can be regarded as fundamental to viral nanotechnology. Functional molecules such as therapeutic or imaging molecules for drug delivery and imaging applications, for example, can be covalently attached and displayed on the VNPs, and this has opened the door for developing "smart" devices for medical applications.

The field of viral nanotechnology is still a young discipline that is rapidly evolving. A broad range of VNP platforms have been exploited or show promise in applications ranging from the development of battery electrodes to medical imaging and drug delivery.

1.3 WHY VNPs? MATERIALS PROPERTIES OF VNPs

When a virologist looks at a virus, he or she might see a pathogen, an infectious agent that is causing a disease. What does a chemist or a materials scientist see in VNPs? Working at the interface of chemistry and medicine, we see tiny building blocks, platforms that can be tuned with functionalities. A VNP can be regarded as a platform that is used as a template or scaffold for the generation of functional materials. The regular surface properties of VNPs allow one to covalently attach functional molecules (termed *functionalizing*). VNPs are *programmable* and *tunable*, as they can be *functionalized* with a broad range of molecules used for manifold applications. The covalent modification can also lead to a *tuning* of the materials properties; for example, the charge properties can be altered by attaching neutral groups to charged surface groups on the viral capsid.

VNPs occur in two basic shapes, icosahedral and rod (Fig. 1.3). An icosahedron is a polyhedron with 20 triangular faces; the icosahedral symmetry description is explained in detail in Chapter 2. More complex structures such as head–tail bacteriophages, enveloped viruses, and even spindle- and bottle-shaped particles can also be found. For example, the particles of *Acidianus* bottle-shaped virus (ABV) indeed look like a bottle

(Fig. 1.2). ABV is an archaeal virus of the family *Ampulliviridae* with particles about 230 nm long and 4–75 nm wide (Prangishvili *et al.*, 2006).

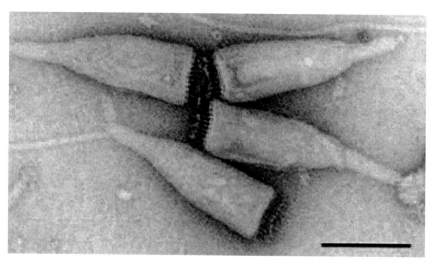

Figure 1.2 Transmission electron micrograph of *Acidianus* bottle-shaped virus. The scale bar represents 100 nm. Reproduced with permission from Prangishvili D., Forterre P., and Garrett R. A (2006) Viruses of the Archaea: a unifying view, *Nat. Rev. Microbiol.*, **4**(11), 837–848.

Icosahedral particles range in their size from 18 nm to 500 nm. Rod-shaped or filamentous VNPs can reach up to 2 μm in length. The structure of many viruses has been solved to atomic or near-atomic resolution, and many of these structures and accompanying structural information can be found at the Virus Particle ExploreR database (VIPER; at www.viperdb.scripps.edu). Structures of VNPs currently under investigation and use for potential applications in nanotechnology are shown in Fig. 1.3. We will discuss each of these systems throughout the book and highlight the characteristics and materials properties of each VNP in Chapter 2.

A whole library of VNPs has become available, offering a large variety of building blocks with varying structural and chemical properties. Each VNP can be selected for its most suitable applications. For example, the reversible permeability of CCMV has led to its use as a constrained chemical reaction vessel (Douglas *et al.*, 2002; Douglas & Young, 1998, 1999). Filamentous or rod-shaped VNPs such as the bacteriophage M13 and TMV have been metal-coated and used as nanowires and nanotubes in the fabrication of lithium–ion battery electrodes and data storage devices (Lee *et al.*, 2009; Nam *et al.*, 2006, 2008; Tseng *et al.*, 2006). The development of VNPs for nanotechnology applications has fueled the search for novel VNPs possessing

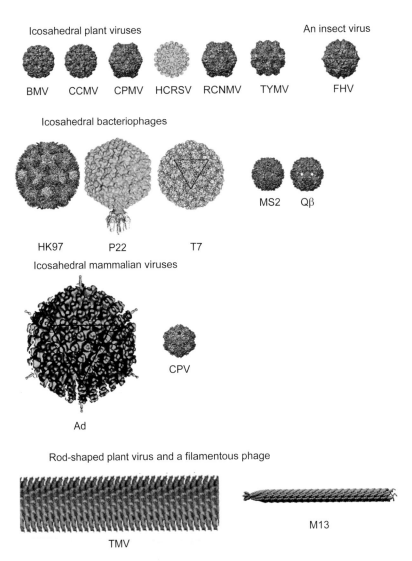

Figure 1.3 A snapshot of the viral nanoparticles (VNPs) that are currently exploited and developed for materials science and medicine. The list is continuously growing; these are the VNPs in use as of May 2009. Icosahedral plant viruses: *Brome mosaic virus* (BMV), *Cowpea chlorotic mottle virus* (CCMV), *Cowpea mosaic virus* (CPMV), *Hibiscus chlorotic ringspot virus* (HCRSV), *Red clover necrotic mottle virus* (RCNMV), *Turnip yellow mosaic virus* (TYMV). Icosahedral insect virus: *Flock house virus* (FHV). Icosahedral bacteriophages: HK97, P22, T7, MS2, and Qβ. Note P22 and T7 are head-tail phages. The tail is shown for P22, not for T7. Icosahedral mammalian viruses: *Adenovirus* and *Canine parvovirus* (CPV). Rod-shaped and filamentous viruses: *Tobacco mosaic virus* (TMV), a plant virus, and M13, a bacteriophage. Images of the following

VNPs were reproduced from the VIPER database (www.viperdb.scripps.edu): BMV, CCMV, CPMV, RCNMV, TYMV, FHV, HK97, MS2, Qβ, and CPV. The structure of HCRSV was reproduced from Doan, D. N., *et al.* (2003) *J. Struct. Biol.*, **144**(3), 253–261. The cryo-electron microscopy structure of P22 was reproduced with permission from Chang, J., *et al.* (2006) *Structure*, **14**(6), 1073–1082. The T7 structure was taken with permission from Agirrezabala , X., *et al.* (2007) *Structure*, **15**, 461–472. *Adenovirus* cryo-electron microscopy reconstruction was reproduced with permission from Johnson, J. E., and Speir J. A. (1997) *J. Mol. Biol.*, **269**(5), 665–675. The cryo-reconstruction of TMV was provided by Bridget Carragher and Clint Potter; data were collected and processed at the National Resource for Automated Molecular Microscopy at the Scripps Research Institute.) M13 was taken with permission from Khalil, A. S., *et al.* (2007) *PNAS*, **104**(12), 4892–4897.

exotic structural or chemical features, and these are often found as infectious agents of extremophile hosts. For example, the hyperthermophile organism *Sulfolobus islandicus* thrives at 80°C and pH < 3.0. The associated filamentous archaeal virus *Sulfolobus islandicus* rod-shaped virus 2 (SIRV2) is both extremely stable owing to its host natural habitat, and uniquely exploitable as a template for site-selective and spatially controlled bioconjugation. Functionalities can be attached and displayed at either the virus body or its ends (Steinmetz *et al.*, 2008).

From a materials science point of view, VNPs are exceptionally robust; as mentioned earlier, a primary function of the capsid is the protection of the encapsidated nucleic acid. As a result temperature- and pH-stability is increased. Several VNPs sustain temperatures as high as 60°C for several hours. In terms of pH stability, various particles remain intact over a pH range of 2–10. These characteristics make them feasible building blocks for the generation of novel materials. Chemists have also found that many VNPs are stable in a range of solvent–buffer mixtures, which is essential for chemical modification procedures (see Chapter 4).

VNPs can be produced on a large scale at low costs and in short time frames (discussed in detail in Chapter 3). The particles have a high degree of symmetry and polyvalency, and they are monodisperse, meaning that every single particle looks virtually identical in size and shape to all other particles formed by that species. In addition to the ability to self-assemble into discrete particles, VNPs also show a propensity for self-organization. Straightforward crystallization procedures lead to self-organization, and 2D and 3D crystals can be readily obtained (Sun *et al.*, 2007; Wang *et al.*, 2002). In addition self-supporting crystalline thin films in the centimeter range, especially of rod-shaped VNPs, can be fabricated (Kuncicky *et al.*, 2006; Lee *et al.*, 2003). (Arrays and films are described in Chapter 7.)

1.4 SUMMARY

A large variety of nanobuilding blocks are available and have commanded the attention of chemists and materials scientists. Initial proof-of-concept studies focused on developing chemical modification strategies. Bioconjugation chemistries that allow for site-selective covalent modification have been adapted to VNPs (discussed in Chapter 4). Further selective entrapment and encapsulation techniques have been developed (see Chapter 5). A broad range of mineralization and metal deposition techniques have also been applied (see Chapter 6). With all these modification protocols in hand, research has moved toward the development of functional devices. Today VNPs are utilized for manifold applications that range from materials (described in Chapters 4 to 7) to biomedicine (discussed in Chapter 8).

Viral nanotechnology is a young discipline just emerging from its infancy. It is an intriguing field with wide-ranging opportunities. It is an exciting time to be working at the virus–chemistry interface!

References

Awram, P., Gardner, R. C., Forster, R. L., and Bellamy, A. R. (2002) The potential of plant viral vectors and transgenic plants for subunit vaccine production, *Adv. Virus Res.*, **58**, 81–124.

Canizares, M. C., Nicholson, L., and Lomonossoff, G. P. (2005) Use of viral vectors for vaccine production in plants, *Immunol. Cell Biol.*, **83**(3), 263–270.

Clark, J. R., March, and J. B. (2006) Bacteriophages and biotechnology: vaccines, gene therapy and antibacterials, *Trends Biotechnol.*, **24**(5), 212–218.

Douglas, T., Strable, E., and Willits, D. (2002) Protein engineering of a viral cage for constrained material synthesis, *Adv. Mater.*, **14**, 415–418.

Douglas, T., and Young, M. (1998) Host-guest encapsulation of materials by assembled virus protein cages, *Nature*, **393**, 152–155.

Douglas, T., and Young, M. (1999) Virus particles as templates for material synthesis, *Adv. Mater.*, **11**, 679–681.

Johnson, J., Lin, T., and Lomonossoff, G. (1997) Presentation of heterologous peptides on plant viruses: genetics, structure, and function, *Annu. Rev. Phytopathol.*, **35**, 67–86.

Kuncicky, D. M., Naik, R. R., and Velev, O. D. (2006) Rapid deposition and long-range alignment of nanocoatings and arrays of electrically conductive wires from Tobacco mosaic virus, . *Small*, **2**(12), 1462–1466.

Lee, S. W., Woods, B. W., and Belcher, A. M. (2003) Chiral smectic C structures of virus-based film, *Langmuir*, **19**, 1592–1598.

Lee, Y. J., Yi, H., Kim, W. J., Kang, K., Yun, D. S., Strano, M. S., Ceder, G., and Belcher, A. M (2009) Fabricating genetically engineered high-power lithium-ion batteries using multiple virus gene, *Science*, **324**(5930), 1051–1055.

Marks, T., and Sharp, R. (2000) Bacteriophages and biotechnology: a review, *J. Chem. Technol. Biotechnol.*, **75**, 6–17.

Nam, K. T., Kim, D. W., Yoo, P. J., Chiang, C. Y., Meethong, N., Hammond, P. T., Chiang, Y M., and Belcher, A. M. (2006) Virus-enabled synthesis and assembly of nanowires for lithium ion battery electrodes, *Science*, **312**(5775), 885–888.

Nam, K. T., Wartena, R., Yoo, P. J., Liau, F. W., Lee, Y. J., Chiang, Y. M., Hammond, P. T., and Belcher, A. M. (2008) Stamped microbattery electrodes based on self-assembled M13 viruses, *Proc. Natl. Acad. Sci. USA*, **105**(45), 17227–17231.

Porta, C., and Lomonossoff, G. P. (1998) Scope for using plant viruses to present epitopes from animal pathogens, *Rev. Med. Vir.*, **8**(1), 25–41.

Prangishvili, D., Forterre, P., and Garrett, R. A. (2006) Viruses of the Archaea: a unifying view. *Nat. Rev. Microbiol.*, **4**(11), 837–848.

Scholthof, H. B., Scholthof, B. G., and Jackson, A. O. (1996) Plant virus vectors for transient expression of foreign proteins in plants, *Annu. Rev. Phytopathol.*, **34**, 229–323.

Shenton, W, Douglas, T., Young, M., Stubbs, G., and Mann, S. (1999) Inorganic-organic nanotube composites from template mineralization of Tobacco mosaic virus,. *Adv. Mater.*, **11**, 253–256.

Steinmetz, N. F., Bize, A., Findlay, K. C., and Lomonossoff, G. P., Manchester, M., Evans, D. J., Prangishvili, D. (2008) Site-specific and spatially controlled addressability of a new viral nanobuilding block: *Sulfolobus islandicus* rod-shaped virus 2, *Adv. Funct. Mater.*, **18**, 3478–3486.

Sun, J., DuFort, C., Daniel, M. C., Murali, A., Chen, C., Gopinath, K., Stein, B., De, M., Rotello, V. M., Holzenburg, A., Kao, C. C., and Dragnea, B. (2007) Core-controlled polymorphism in virus-like particles. *Proc. Natl. Acad. Sci. USA*, **104**(4), 1354–1359.

Tseng, R. J., Tsai, C., Ma, L., Ouyang, J., Ozkan, C. S., and Yang, Y. (2006) Digital memory device based on Tobacco mosaic virus conjugated with nanoparticles, *Nat. Nanotechnol.*, **1**, 72–77.

Wang, Q., Lin, T., Tang, L., Johnson, J. E., and Finn, M G. (2002) Icosahedral virus particles as addressable nanoscale building blocks, *Angew. Chem. Int. Ed.*, **41**(3), 459–462.

Woese, C. R., and Fox, G. E. (1977) Phylogenetic structure of the prokaryotic domain: the primary kingdoms. *Proc. Natl. Acad. Sci. USA*, **74**(11), 5088–5090.

Zaitlin, M. (1898) The discovery of the causal agent of the tobacco mosaic disease, in *Discoveries in Plant Biology* (ed. Kung, S. D., and Yang, S. F.), World Publishing, Hong Kong, pp. 105–110.

Chapter 2

OVERVIEW OF THE MANIFOLD VNPs USED IN NANOTECHNOLOGY

This chapter will provide an overview of the different viral nanoparticle (VNP) building blocks that are currently in use for nanotechnology. Some of these platforms have been extensively studied and used for various applications. Other particles are relatively new to the field. The list of VNPs studied for nanotechnology is constantly expanding. As more and more researchers become interested in the manifold potentials of VNPs, the field is rapidly evolving. Studying new VNPs and their unique properties has led to novel ideas and concepts for VNP fabrication and utility.

2.1 A GENERAL INTRODUCTION TO VIRUS STRUCTURE AND REPLICATION

Viruses are ubiquitous in nature; they infect all forms of life. In contrast to other microbes, viruses do not have a cellular metabolism by themselves; rather, they are obligate intracellular parasites, that is, they need the host cell for replication. Viruses may be generally regarded as non-free-living organisms.

In comparison to a cellular organism, viruses generally have a more simple structural and genetic organization. The term virion describes the complete virus particle. A virion consists of nucleic acid and a capsid or other structure to encapsidate the nucleic acid. Some viruses have an additional lipid, or envelope, component. The envelope is a portion of the host cell membrane that is gained during egress and escape of virions from the host cell. Viruses with such an envelope are termed enveloped viruses.

The capsid refers to the protective protein coat that encapsulates and protects the nucleic acid. Virus genomes encode all the information required to facilitate the particle proceeding through its life cycle (for a general

Viral Nanoparticles: *Tools for Materials Science and Biomedicine*
By Nicole F. Steinmetz and Marianne Manchester
Copyright © 2011 by Pan Stanford Publishing Pte. Ltd.
www.panstanford.com

textbook on virology the reader is referred to *Fundamental Virology*, Knipe & Howley, 2001). The general steps of virus replication are as follows:

1. **Attachment and penetration:** In general, animal viruses, insect viruses, bacteriophages, and archaeal viruses bind to specific receptors on the host cellular surface, which induces penetration or internalization of the virus into the host cell. Plant viruses, in general, do not recognize specific cell surface receptors; rather, they enter the cells through lesions that are, for example, those caused by feeding insects.

2. **Uncoating:** Unpacking of the nucleic acids from the capsid and initiation of a new replication cycle.

3. **Replication and protein expression:** Copies of the nucleic acids are synthesized and coat proteins, as well as non-structural proteins, are produced. The coat proteins then self-assemble into the capsid and package the nucleic acid.

4. **Assembly:** Capsid assembly and maturation, and specific packing of the nucleic acid into the capsid — generally performed with high precision. Studying and understanding virus assembly is an important goal in virology, structural biology, and nanotechnology. The self-assembly process of viruses has been exploited in nanotechnology and is discussed in Chapter 5.

5. **Escape:** The assembled particles are released from the infected cell, either by cell lysis or by budding (the latter is typically the case for enveloped viruses). The virus derives its envelope from the host cell membrane during this step.

6. **Transfer:** Survival of transfer from the infected cell to the next target cell. Transfer refers to cell-to-cell movement, as well as the transfer from one host to another. Transfer between hosts is achieved in various ways; some viruses are transferred by insect vectors, whereas others are transferred from host to host via body fluids or aerosols.

2.1.1 The Virus Genome

Virus genomes are highly diverse in their nucleic acid composition. The virus genome can be composed of either DNA or RNA. It can be linear, circular, or segmented. The nucleic acid can be single-stranded or double-stranded. Single-stranded genomes can be positive or negative sense, where the *sense* refers to the messenger RNA (mRNA) polarity of the nucleic acid strands. Positive sense refers to the RNA sequence that can be directly translated into a protein; it can also be called the coding strand. Negative sense is the complementary strand; it is also referred to as antisense and requires transcription into the positive-sense strand prior to translation into a protein.

For detailed information about the genome organization and structure of viruses see the following databases: Description of Plant Viruses database (DPV; http://www.dpvweb.net), the Universal Database of the International Committee on Taxonomy on Viruses (ICTV; http://www.ncbi.nlm.nih.gov/ICTVdb), and the Virus Particle ExploreR Database (VIPER; http://viperdb.scripps.edu).

Chimeric virus strategies used in nanotechnology. Because virus genomes are relatively small, many virus genomes have been completely sequenced, and detailed knowledge about their genomic properties is available (nucleic acid sequences are available at the National Center for Biotechnology Information database; http://www.ncbi.nlm.nih.gov/.) This led to the development of chimeric virus technology. A virus chimera refers to a genetically modified version of the wild-type or native virus. Chimeric viral particles, for example, have been used for vaccine development. Chimeric virus technology allows genetic insertion of an antigenic peptide into the coat protein sequence. Viral capsids consist of multiple copies of the same coat protein; hence, the chimeric particle displays multiple copies of the antigenic peptide sequence on its capsid surface. The multivalency of peptide presentation and the VNP carrier may lead to an enhanced immune response against the antigens. Vaccine strategies are discussed in Chapter 8.

In nanotechnology, chimeric virus technology is also used to fine-tune and alter VNP surface properties. Additional amino acid side chains can be introduced onto the capsid to allow chemical modification and the installation of additional functionalities. For example, if a VNP does not have any solvent-exposed Cys side chains on its exterior capsid surface, these could be introduced using molecular cloning techniques. The thiol group of Cys residues is an attractive target for bioconjugation chemistry because thiols undergo facile coupling with maleimide-activated compounds, and a large range of such compounds are commercially available. A range of Cys-added chimeric VNPs have been synthesized, which will be mentioned throughout this book. The principles of chimeric virus technology and its applications for nanotechnology are discussed in greater detail in Chapter 3.

2.1.2 The Structure of Viruses

The structures of viruses reveal remarkable diversity (recall the complex and unique structure of the *Acidianus* bottle-shaped virus ABV in Chapter 1; Fig. 1.2). The structure of viruses can be divided into four main groups:

1. Icosahedral
2. Rod-shaped: helical tubes or filaments
3. Enveloped
4. Complex

Examples of each group are shown in Fig. 2.1. Many viruses currently in use for nanotechnology have icosahedral symmetry (see Section 2.1.2.1). Approximately 20 different viruses are currently being studied and exploited as VNPs for nanotechnology, 13 of which are non-enveloped with icosahedral symmetry. Four are non-enveloped rod-shaped, and three are enveloped.

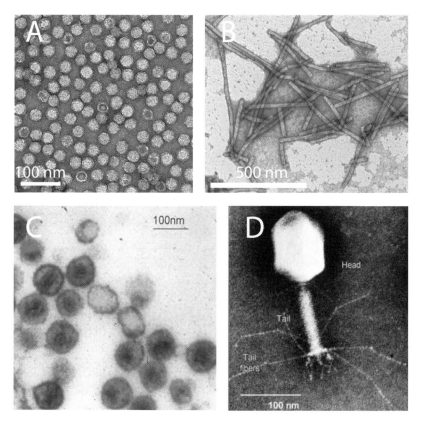

Figure 2.1 Transmission electron micrographs of viruses illustrating the different structures. (A) *Cowpea mosaic virus* particles as an example of icosahedral particles. (B) *Tobacco mosaic virus* particles represent rod-shaped particles. (C) To illustrate an enveloped virus *Human immunodeficiency virus* 1 is shown. (D) Bacteriophage T4 has a complex head and tail structure. Images were reproduced from Steinmetz, N. F. (2007) Viral capsids as programmable nanobuilding blocks, *PhD thesis*, John Innes Centre, University of East Anglia, Norwich (A, B), and the universal database of the International Committee on Taxonomy on Viruses (ICTV; C, D).

Some reasons for the lowered incidence of enveloped VNPs in the nanotech literature are as follows: (i) production of non-enveloped VNPs are more feasible compared with generating enveloped VNPs; (ii) non-enveloped VNPs are in general more stable; and (iii) use of non-enveloped non-mammalian VNPs can be regarded as safe from a human health perspective compared to working with potentially infectious enveloped mammalian VNPs.

Studies concerning the use of enveloped viruses for nanotechnology have utilized *Rubella virus* (RV) and *Influenza virus* A/PR8 (H1N1) (Fischlechner *et al.*, 2005, 2006, 2007; Toellner *et al.*, 2006). The envelope was exploited in binding the particles onto solid supports for sensing applications. These devices may prove useful in viral diagnostics (discussed in Chapter 7). Another enveloped VNP that has been studied is *Chilo irisdescent virus* (CIV) (Radloff *et al.*, 2005). However, the envelope was removed prior to studying the particle for potential applications in nanotechnology.

2.1.2.1 Icosahedral particles

Description of icosahedral symmetry. The word *icosahedron* derives from the Greek language and means "twenty seat." An icosahedron is a polyhedron with 20 triangular faces. It has fivefold, threefold, and twofold rotational symmetry axes, in short: 5:3:2 symmetry (Fig. 2.2).

Figure 2.2 Schematic of an icosahedron with 20 triangular faces and 5:3:2 symmetry. The fivefold axis is shown as red pentagrams, the threefold axis is depicted as a blue triangle, and the twofold axis is highlighted by green rectangle.

The icosahedron as a structure is the largest closed shape that can be formed by identical units (the coat proteins). Icosahedral symmetry requires a defined number of coat protein units. To form an icosahedral virus particle with 5:3:2 symmetry 60 coat protein subunits are required. Three coat proteins are placed on each of the 20 triangular faces in an equivalent manner (3 subunits × 20 faces = 60 coat protein subunits) (Caspar & Klug,

1962; Johnson & Speir, 1997). There are a few viruses whose capsids indeed consist of 60 coat protein subunits, for example, the mammalian virus *Canine parvovirus* (CPV) (2.2.15) (see the VIPER; http://viperdb.scripps.edu).

However, most viruses form larger structures and consist of more than 60 coat protein subunits. Particles that are formed by more than 60 units are described to have quasi-equivalent symmetry because the subunits cannot be placed at equivalent positions (they are placed at quasi-equivalent positions). To assemble quasi-equivalent particles, conformational switching of the subunits is required (Johnson & Speir, 1997). The geometric design principles for quasi-equivalence of larger virus capsids were developed by Caspar and Klug in 1962 when they introduced the concept of *triangulation (T) numbers* (Caspar & Klug, 1962).

The triangulation (*T*) numbers. According to Caspar and Klug's theory (Caspar & Klug, 1962), the icosahedral virus capsid consists of pentamers and hexamers. A virus particle looks pretty much like a football (soccer ball). A football can be regarded as a spherically truncated icosahedron consisting of pentagons and hexagons (Fig. 2.3).

Figure 2.3 Schematic of a football with pentagons (black) and hexagons (white).

In the viral capsid, the same coat protein forms the pentamers and hexamers; the bonding relation and their environment are thus not identical. This distortion is called quasi-equivalence. Pentamers are inserted in place of certain hexamers, in accordance with selection rules described by the *T* number. If we assume a flat sheet of hexamers (Fig. 2.4), the relative position of the hexamers can be indexed along the axis denoted by *h* and *k* related by a 60° rotation. The mathematical relation is given in the following formula (Caspar & Klug, 1962):

$$T = h^2 + hk + k^2 \qquad (2.1)$$

with *h* and *k* being any positive integer or zero. *T* numbers can thus only adopt positive integer values. The size of the capsid is proportional to the *T* number. The larger the *T* number, the larger the capsid (Fig. 2.5).

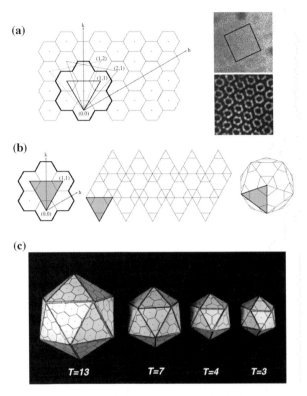

Figure 2.4 Geometric principles for generating icosahedral quasi-equivalent surface lattices. These constructions show the relation between icosahedral symmetry axes and quasi-equivalent symmetry axes. The latter are symmetry elements that hold only in a local environment. (A) Hexamers are initially considered planar, and pentamers are considered convex, introducing curvature in the sheet of hexamers when they are inserted. The closed icosahedral shell, composed of hexamers and pentamers, is generated by inserting 12 pentamers at appropriate positions in the hexamer net. To construct a model of a particular quasi-equivalent lattice, one face of an icosahedron is generated in the hexagon net. The origin is replaced by a pentamer, and the (h,k) hexamer is replaced by a pentamer. The third replaced hexamer is identified by threefold symmetry (i.e., complete the equilateral triangle of the face). (b). Seven hexamer units (bold outlines in (a)) defined by the $T = 3$ lattice choice are shown, and the $T = 3$ icosahedral face defined in (a) has been shaded. A three-dimensional model of the lattice can be generated by arranging 20 identical faces of the icosahedron as shown, and folded into a quasi-equivalent icosahedron. (c) Cardboard models of several icosahedral quasi-equivalent surface lattices constructed using the method described above. The procedure for generating quasi-equivalent models described here does not exactly correspond to that the described by Caspar & Klug (1962); however, the final models are identical with those described in their paper. Reproduced with permission from Johnson, J. E., and Speir, J. A. (1997) Quasi-equivalent viruses: a paradigm for protein assemblies, *J. Mol. Biol.*, 269(5), 665–675.

Many viruses have $T = 3$ icosahedral symmetry. These include *Brome mosaic virus* (BMV), *Cowpea chlorotic mottle virus* (CCMV), *Hibiscus chlorotic ringspot virus* (HCRSV), *Red clover necrotic mottle virus* (RCNMV), *Turnip yellow mosaic virus* (TYMV), *Flock House virus* (FHV), MS2, and Qβ. All of these particles are formed by 180 identical coat protein subunits (60 T subunits). There is, however, a wide range; the capsid of P22 has icosahedral $T = 4$ symmetry, T7 and HK97 have $T = 7$ symmetry, and *Adenovirus* (Ad) has $T = 25$ symmetry. The enveloped virus *Rubella* has capsids of $T = 4$ symmetry. The largest virus utilized for nanotechnology is the enveloped virus CIV, and its capsids have $T = 147$ symmetry. Structural information including T values can be found using the databases: VIPER; http://viperdb.scripps. edu, and ICTV; http://www.ncbi.nlm.nih.gov/ICTVdb. To illustrate the different symmetries cryo-electron microscopy reconstruction images of CCMV, HK97, and Ads are given in Fig. 2.5. In theory, any size of icosahedron can be formed. However, the largest isometric particle formed of a single subunit type found in nature has $T = 7$ symmetry. If larger structures are formed, the capsids appear to require one or more additional structural scaffolding proteins that control and stabilize the self-assembly process (Johnson & Speir, 1997). This is certainly the case for Ad ($T = 25$), in which the capsid is formed with 12 additional structural proteins (see Section 2.2.12) (Rux & Burnett, 2004; San Martin & Burnett, 2003).

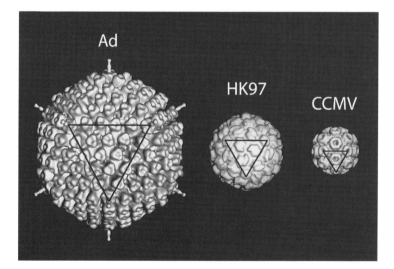

Figure 2.5 Cryo-electron microscopy reconstruction images of the mammalian virus *Adenovirus* (Ad) ($T = 25$), the bacteriophage HK97 ($T = 7$), and the plant virus *Cowpea chlorotic mottle virus* ($T = 3$). The triangular coordinates are shown as black triangles. Reproduced with permission from Johnson, J., Lin, T., and Lomonossoff, G. (1997) Presentation of heterologous peptides on plant viruses: genetics, structure, and function,. *Annu. Rev. Phytopathol.*, **35**, 67–86.

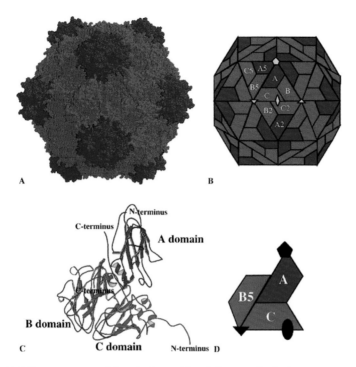

Figure 2.6 The structure of the viral capsid and the icosahedral asymmetric unit of *Cowpea mosaic virus* (CPMV). The capsid is comprised of two viral proteins, the S and L subunits, which form three β-sandwich domains in the icosahedral asymmetric unit. These three domains occupy comparable positions shown in a $T = 3$ quasi-equivalent surface lattice. These are designated A (domains forming pentamers) and B and C (domains forming hexamers). Numerals following the letter indicate positions related by the icosahedral symmetry to A, B, and C. The S subunit forms the domain occupying the A positions in an icosahedral lattice and these are shown in blue. The L subunit is formed by two domains, C and B5, shown in green and red, respectively. (A). A space-filling drawing of the CPMV capsid. All atoms are shown as spheres of 1.8 Å in diameter. (B). A schematic presentation of the CPMV capsid. The S subunit occupies A positions around the fivefold axis; the two domains of L subunit occupy the B and C positions. The quaternary interactions at A/B5 and C/C2 interfaces are pseudo-equivalent. Selected icosahedral symmetry axes are also shown. (C). A ribbon diagram of the three β-barrel domains that comprise the icosahedral asymmetric unit. N-termini of the S and L subunits are in the interior, and C-termini are in the exterior. All three domains are variants of canonical jelly-roll β sandwiches. (D). A schematic diagram of the asymmetric unit with icosahedral symmetry axes. The A domains surround the fivefold axes, and the B and C domains alternate around the threefold axis. Reproduced with permission from Lin, T., Chen, Z., Usha, R., Stauffacher, C. V., Dai, J. B., Schmidt, T., and Johnson, J. E. (1999) The refined crystal structure of *Cowpea mosaic virus* at 2.8 Å resolution, *Virology*, **265**(1), 20–34.

The *P* numbers. Most virus capsids consist of identical copies of coat proteins. However, there are also capsids that are formed by several different coat proteins. In *Cowpea mosaic virus* (CPMV) two different coat proteins form the capsid, the small (S) and large (L) coat protein subunits. The S protein is a one-domain (A domain) and the L protein is a two-domain protein (B and C domains). The three domains can be arranged in so-called *pseudo-equivalence*. The *T* numbers are assigned as *pseudo T* or *P* numbers. CPMV has *P* = 3 symmetry and consists of 60 copies of each coat protein subunit (Fig. 2.6; Lin, T., Chen, Z., Usha, R., Stauffacher, C. V., Dai, J. B., Schmidt, T., and Johnson, J. E. (1999) The refined crystal structure of cowpea mosaic virus at 2.8 Å resolution, *Virology*, **265**(1), 20–34.).

Tuning capsid symmetry using nanotechnology. Nanotechnology has been applied as a technique to understand virus assembly. In particular, synthetic nanoparticles can be used as cores to nucleate assembly of BMV particles (Chen *et al.*, 2006; Dixit *et al.*, 2006; Dragnea *et al.*, 2003; Sun *et al.*, 2007) (see Chapter 5). Assessing whether the morphology of the capsid could be tuned using different-sized nanoparticle cores, it was found that the capsid symmetry indeed is dependent on the nanoparticle core size. Incorporated gold cores of 9 nm yielded particles similar to *pseudo T* = 2 symmetry,[1] particles with gold cores of 6 nm size showed *T* = 1 symmetry, and particles with cores of a size of 12 nm resulted in particles similar to native particles with *T* = 3 symmetry (Sun *et al.*, 2007). Studies along these lines are expected to provide further insights into the fundamental understanding of viral self-assembly.

2.1.2.2 Rod-shaped VNPs: helical tubes and filamentous structures

Icosahedral particles are relatively more common than rod-shaped or filamentous structures. There are, however, a number of rod-shaped particles relevant to nanotechnology including: *Tobacco mosaic virus* (TMV), *Potato virus X* (PVX), M13, and *Sulfolobus islandicus* rod-shaped virus 2 (SIRV2) (for example, see references: Carette *et al.*, 2007; Lee *et al.*, 2009; Mao *et al.*, 2004; Nam *et al.*, 2006, 2008; Schlick *et al.*, 2005; Shenton *et al.*, 1999; Steinmetz *et al.*, 2008; Tseng *et al.*, 2006; Yi *et al.*, 2007; Yoo *et al.*, 2006).

---

[1] *T* = 2 symmetry is typically referred to as *pseudo T* = 2 symmetry. According to the model by Kasper and Klug (see Eq. 1), a *T* = 2 would not be feasible. However, proteins and potein–protein interactions are flexible — also allowing symmetries that would not have been predicted using mathematical models.

Helical structures are formed by multiple copies of coat protein stacked around a central axis. The nucleic acid binds the protein helix via electrostatic interactions. The properties of the helical structure can be described by the number of turns, the number of coat proteins per turn, and the *pitch*, meaning the distance along the helical axis corresponding to exactly one turn (for a textbook on virus structure and assembly see *Advances in Virus Research* by Roy, 2005). TMV, for example, is about 300 nm in length, has a diameter of 18 nm, and is formed by 2130 identical coat proteins. These coat proteins form a closely packed helix with a *pitch* of ca. 2.3 nm with 16 1/3 subunits per turn. The tube has a hollow cylindrical channel of about 4 nm in diameter (Namba & Stubbs, 1986) (see also DPV; http://www.dpvweb.net, and the ICTV; http://www.ncbi.nlm.nih.gov/ICTVdb) (Fig. 2.7). The ends of the rods can sometimes comprise additional structural proteins; this is the case for the filamentous particles M13 and SIRV2 (discussed in detail in Section 2.3.3 and 2.3.4).

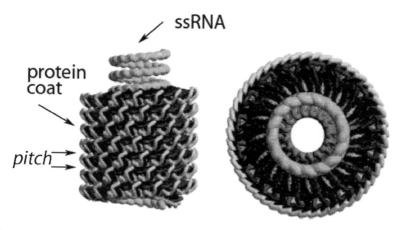

Figure 2.7 Schematic of the structure of the rod-shaped virus *Tobacco mosaic virus.* On the left a view perpendicular to the helix; on the right a view down to the helix axis. Reproduced with permission from Roy, P. (2005) *Advances in Virus Research*, Elsevier Academic Press.

The structures of viral rods can be short or long, and rigid or flexible. For example, TMV particles are 300 nm in length and form rigid rods. In contrast, M13 is nearly 1 μm in length and a highly flexible structure. Indeed, it is so flexible that its ends can be linked together to generate nanorings (Nam *et al.*, 2004).

2.1.2.3 Enveloped viruses

Enveloped viruses are common among animal viruses. *Rubella* virus (RV), for one, is an enveloped virus utilized in nanotechnology. The capsid structure is icosahedral with $T = 4$ symmetry. A lipid bilayer forms an additional structural layer (the envelope) around the capsid (Frey, 1994). The envelope is a piece of the host cell lipid membrane that the particles acquire when they exit the cell by a mechanism termed budding.

To date, few enveloped viruses have been utilized in nanotechnology, primarily because generating the quantities of material required is more feasible using non-enveloped viruses, and because non-enveloped capsids are generally more stable. Nevertheless, efforts have been made using enveloped VNPs for the development of novel nanoscale diagnostic devices (Fischlechner *et al.*, 2005, 2006, 2007; Toellner *et al.*, 2006) (discussed in Chapter 7).

2.1.2.4 Complex structures

Many bacteriophages have complex structures. For example, particles of the bacteriophages HK97, P22, and T7 consist of an icosahedral head (capsid) coupled to a tail. The tail is a hollow structure through which the bacteriophage's nucleic acid is injected into the host cell after attachment. The tail structure can be large and complex, or small and simple.

For HK97, P22 (Kang *et al.*, 2008), and T7 (Liu *et al.*, 2005, 2006) the capsids alone have been utilized as platforms in nanotechnology. Besides the capsids, researchers have also undertaken approaches in which components of the tail assembly of the phage T4 were employed. The tail of T4 is complex in that it consists of both a hollow tube and an additional cup structure (Fig. 2.8). The cup structure is the connecting portion between the tube and the capsid. Genetic and chemical engineering of the cup structure allowed the accommodation of gold nanoclusters and metal complexes within the cup-shaped space (Koshiyama *et al.*, 2008; Ueno *et al.*, 2006). These heterostructures may find applications as oxidative catalysts.

2.2 ICOSAHEDRAL VNPs

2.2.1 *Brome Mosaic Virus*: A Plant Virus

BMV is a plant virus from the family *Bromoviridae*. It infects about 60 genera within the *Gramineae*, including maize and barley plants, in which it causes

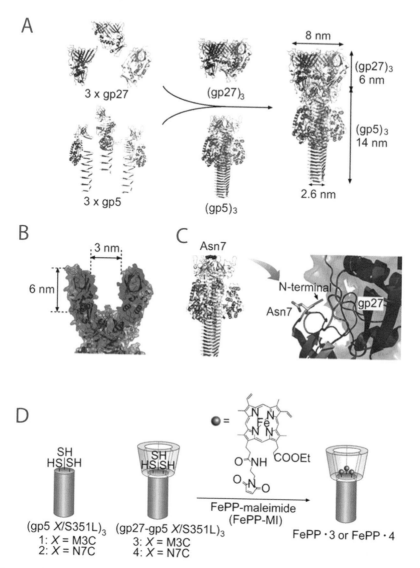

Figure 2.8 (A). The crystal structure of the bionanocup structure from bacteriophage T4. (B). The structure is formed by three copies each of gene product gp27 and gp5 (gp27-gp5)₃. (C). Section view of the bionanocup, and introduction sites of (gp5)₃ for Cys residues. Asn residues were exchanged with Cys side chains to allow for chemical modification. (D). Schematic drawing of the conjugation reactions of FePPmaleimides (= catalytic metal complexes) and (gp5)₃ or (gp27-gp5)₃ cysteine mutants. Reproduced with permission from Koshiyama, T., Yokoi, N., Ueno, T., Kanamaru, S., Nagano, S., Shiro, Y., Arisaka, F., and Watanabe, Y. (2008) Molecular design of heteroprotein assemblies providing a bionanocup as a chemical reactor, *Small*, **4**(1), 50–54.

mild mosaic symptoms. BMV is transmitted by nematodes and can be found in the USA and in Europe (from DPV; http://www.dpvweb.net and the ICTV; http://www.ncbi.nlm.nih.gov/ICTVdb).

BMV particles have icosahedral symmetry with a diameter of about 28 nm; the structure has been solved by X-ray crystallography (Lucas *et al.*, 2002). The capsid is composed of 180 identical copies of a single coat protein arranged in $T = 3$ symmetry. The genome is positive-sense, single-stranded, and tripartite, meaning that the genetic information is encoded on three different RNA molecules (Ahlquist *et al.*, 1981, 1984).

BMV particles are typically propagated in barley plants (*Hordeum vulgare*). Yields in gram scales can be obtained from 1 kg of infected leaf material. A heterologous expression system in yeast (*Saccharomyces cerevisiae*) has also been developed that allows high yield production of virus-like particles (VLP) (reviewed in Schneemann & Young, 2003). A VLP is a particle that is produced in a non-natural host. VLPs do not contain genomic information; the particles either are empty or pack random cellular RNA. VLPs are of interest in nanotechnology because they do not contain genetic information, and they are non-infectious and considered safe from an agricultural and human health perspective. The different expression systems and their advantages will be described further in Chapter 3.

Besides the ability to produce empty BMV VLPs in a heterologous expression system, coat protein monomers can also be assembled *in vitro* into intact VLPs (Pfeiffer *et al.*, 1976; Pfeiffer & Hirth, 1974). These characteristics have led to the development of BMV as a platform for the encapsulation of synthetic materials such as gold nanoparticles and quantum dots (QDs) (Chen *et al.*, 2006; Dixit *et al.*, 2006; Dragnea *et al.*, 2003; Sun *et al.*, 2007) (described earlier in this chapter and discussed in detail in Chapter 5).

2.2.2 *Cowpea Chlorotic Mottle Virus*: A Plant Virus

CCMV, like BMV, belongs to the family *Bromoviridae*, which is reflected by a range of similarities in structure and genome organization. CCMV infects mostly legumes and causes disease in black-eyed peas as well as in soybeans. The virus is found exclusively in the United States of America where it is transmitted by beetles (from DPV; http://www.dpvweb.net).

CCMV forms particles with a diameter of about 30 nm. The CCMV capsid is composed of 180 identical copies of a single coat protein arranged in a $T = 3$ symmetry. The structure of the particles has been determined by X-ray crystallography (Speir *et al.*, 1995). The genome is a tripartite, positive-sense, single-stranded RNA genome.

Particle production can be achieved in natural hosts, black-eyed peas (in Latin *Vigna unguiculata*), or heterologous expression systems. In the natural host, high titers of VNPs accumulate and 1–2 g can be isolated from 1 kg of infected leaf material. Heterologous expression systems give rise to comparably high yields; for example, expression of CCMV VLPs in yeast (*Pichia pastoris*) yields up to 0.5 g VLPs per kg wet cell mass(Brumfield *et al.*, 2004). Further coat protein monomers can be expressed in *Escherichia coli* and then be self-assembled *in vitro* into intact empty CCMV protein cages (Zhao *et al.*, 1995).

The structural transitions of CCMV. CCMV particles are highly dynamic platforms. CCMV particles undergo reversible pH- and metal ion-dependent structural transitions. These structural transitions are described as a swelling mechanism, which results in an approximately 10% increase in the particle dimension. The structural transition is a result of an expansion of the threefold axis of the virus particle, and the swelling is accompanied by the formation of 60 separate 2-nm-sized openings in the protein shell. Under swollen condition these openings allow free molecular exchange between the virus cavity and the surrounding bulk medium (Fig. 2.9) (Liepold *et al.*, 2005; Speir *et al.*, 1995).

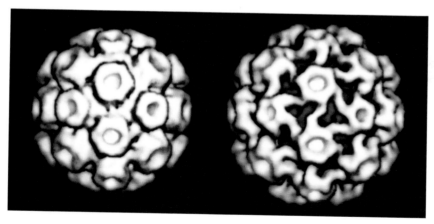

Figure 2.9 Cryo-electron microscopy and image reconstruction of *Cowpea chlorotic mottle virus*. "Closed" conformation (pH ≤ 6.5, metal ions) (on the left), and "open" conformation (pH ≥ 6.5, no metal ions present) (on the right). Reproduced with permission from Liepold, L. O., Revis, J., Allen, M., Oltrogge, L., Young, M., and Douglas, T. (2005) Structural transitions in *Cowpea chlorotic mottle virus* (CCMV), *Phys. Biol.*, **2**(4), 166–172.

The dynamics and reversible swelling of the CCMV capsid have been exploited for nanotechnology and is also referred to as *gating*. In the swollen form there is free exchange between the surrounding medium or buffer and the interior of the viral capsid. Molecules smaller than 2 nm (that is the diameter of the pores) can diffuse into the CCMV capsid. Structural transition to the closed form leads to entrapment of the material within the capsid; the molecules are *gated*.

The dynamic properties of the CCMV capsid makes it an attractive building block for applications ranging from materials to medicine. On the materials side the capsids can be exploited to selectively and reversibly entrap materials and allow spatially controlled synthesis of monodisperse inorganic particles that may be difficult to obtain using common synthetic procedures (see Chapters 5 and 6). With a viewpoint on medical applications, the reversible gating may be exploited for the entrapment and controlled release of therapeutic molecules. Not only the dynamic gating mechanism of the particles is intriguing, using *in vitro* assembly methods, it has been shown that the CCMV coat protein facilitates the construction of a range of structures including tubes, rosettes, and sheets (Bancroft *et al.*, 1969).

2.2.3 *Cowpea Mosaic Virus*: A Plant Virus

CPMV is a member of the Comovirus genus. Members are also known as plant picorna-like viruses because they share similarities in structure, genome organization, and replication strategy with animal picornaviruses (for example *Poliovirus*, the infectious agent responsible for the human disease poliomyelitis).

CPMV has a rather narrow natural host range; it infects legumes and was first reported in black-eyed peas (*V. unguiculata*), which is also the plant used for propagation of the particles. Geographically CPMV is found in Cuba, Japan, Kenya, Nigeria, Surinam, Tanzania, and the USA, where the virus is transmitted by leaf-feeding beetles, thrips, and grasshoppers (from DPV; http://www.dpvweb.net and the ICTV; http://www.ncbi.nlm.nih.gov/ICTVdb).

The virions of CPMV are approximately 30 nm in size and are formed by 60 copies of two different types of coat proteins: S (one domain) and L (two domains). The three domains of the two coat proteins form the asymmetric unit and are arranged in a similar surface lattice to $T = 3$ viruses, except that they have different polypeptide sequences; therefore, the particle structure is described as $P = 3$ symmetry (Rossmann & Johnson, 1989). CPMV has a bipartite positive-sense, single-stranded RNA genome (Lomonossoff & Shanks, 1983; van Wezenbeek *et al.*, 1983).

CPMV wild-type and mutant particles are propagated in black-eyed peas, in which the particles accumulate to high titers (1–2 g/kg infected leaf material). In addition to the natural hosts, species from several families —

including legumes and *Nicotiana benthamiana* — are known to be susceptible to the virus and transmission can be achieved experimentally by mechanical inoculation (Lomonossoff & Shanks, 1999). Heterologous expression of CPMV VLPs using baculovirus expression vectors (an insect virus expression system that is discussed in detail in Chapter 3) has been achieved (Saunders *et al.*, 2009; Shanks & Lomonossoff, 2000). Expression of CPMV VLPs is currently under development. The research team led by Lomonossoff (John Innes Centre, Norwich, UK) is currently optimizing heterologous expression of CPMV VLPs using the baculovirus system. They also developed a novel system that allows expression of CPMV coat proteins and assembly of VLPs *in planta* (Saunders *et al.*, 2009).

T, M, and B components of CPMV. CPMV particles can be separated by density into three components, which have identical protein composition but differ in their RNA contents (Bancroft, 1962; Bruening & Agrawal, 1967; Wu & Bruening, 1971). The particles of the top (T) component are devoid of RNA, whereas the M and B components each contain a single RNA molecule, RNA-2 and RNA-1, respectively (Lomonossoff & Johnson, 1991) (Fig. 2.10). T components can be isolated and are of potential interest for nanotechnology. Empty particles are not infectious and thus a desired starting material for the development of functional devices.

Top component
devoid of RNA

Middle component
contains RNA-2

Bottom component
contains RNA-1

Figure 2.10 Separation of the different nucleocomponents of *Cowpea mosaic virus* particles in a Nycodenz density gradient by ultracentrifugation. The bands of the top (T), middle (M) and B (bottom) component are visible in the gradient. Reproduced with permission from Steinmetz, N. F., Evans, D. J., and Lomonossoff, G. P. (2007) Chemical introduction of reactive thiols into a viral nanoscaffold: a method that avoids virus aggregation, *ChemBioChem*, **8**(10), 1131–1136.

The applications of CPMV are manifold, including presentation platforms, building blocks for the construction of multilayered arrays, and imaging tools for medicine. For the icosahedral viruses, CPMV is one of the most extensively studied virions in the field of viral nanotechnology and will be discussed in all chapters of this book.

2.2.4 *Hibiscus Chlorotic Ringspot Virus*: A Plant Virus

HCRSV is a virus in the genus Carmovirus. It exists throughout the world and infects plants of the family *Malvaceae*. From an agricultural point of view, HCRSV has limited economic importance. It mainly causes disease in *Hibiscus rosa-sinensis* cultivars; symptoms are mild and plants continue to flower.

The particles are about 30 nm in diameter and formed by 180 copies of identical coat proteins; thus, the particles have icosahedral $T = 3$ symmetry. The genome is a positive-sense, single-stranded RNA genome (from DPV; http://www.dpvweb.net).

HCRSV is relatively new to the field for nanotechnology and only few studies have been described. Here, the VNPs were exploited for an *in vitro* targeted drug-delivery approach (Ren *et al.*, 2006, 2007); see Chapter 8.

2.2.5 *Red Clover Necrotic Mottle Virus*: A Plant Virus

RCNMV is from the family *Tombusviridae*. The virus can naturally be found in red clover in Europe. More than 100 plants can be infected including cucumber, tobacco, basil, some beans, and peas (from DPV; http://www.dpvweb.net).

RCNMV virions contain 180 identical coat protein copies arranged in an icosahedral $T = 3$ symmetry; the capsids have a diameter of approximately 36 nm (Sherman *et al.*, 2006). The genome is bipartite positive-sense, single-stranded RNA (Basnayake *et al.*, 2006).

Nicotiana clevelandii and *Phaseolus vulgaris* are typically used for virus propagation. The self-assembly process of the RCNMV capsid is well understood, and *in vitro* assembly protocols have been established. The self-assembly of the coat proteins is initiated and stabilized by an internal protein/RNA cage known as the origin of assembly site (Sit *et al.*, 1998). This knowledge has been exploited for nanotechnology. Synthetic nanoparticles such as gold nanoparticles can be incorporated into RCNMV capsids by creating an artificial origin of assembly site (OAS) on the nanoparticle, thus initiating the *in vitro* assembly and the formation of VNPs around the nanoparticle cores (Loo *et al.*, 2006, 2007) (see Chapter 5).

2.2.6 *Turnip Yellow Mosaic Virus*: A Plant Virus

TYMV is from the family *Tymoviridae*. The virus can be found in Europe. The host range is confined to plants from the family *Cruciferae* also known as the mustard or cabbage family. Chinese cabbage is used for propagation. TYMV virions have icosahedral $T = 3$ symmetry; they consist of 180 identical coat protein units, and are about 28 nm in diameter. The genome is positive-sense, single-stranded RNA (from DPV; http://www.dpvweb.net).

2.2.7 *Flock House Virus*: An Insect Virus

FHV is a member of the *Nodaviridae* family. It infects insects, specifically the New Zealand grass grub. The virus was named after a former agricultural research station (Flock House) on the north island of New Zealand, where scientists discovered that FHV was responsible for dead grass grubs.

FHV particles consists of 180 identical coat proteins arranged in a $T = 3$ symmetry lattice; the structure has been solved by X-ray crystallography (Fisher *et al.*, 1993). The genome is a bipartite, positive-sense, single-stranded RNA genome (Krishna & Schneemann, 1999).

FHV particles can be produced in *Drosophila* cells with yields of 1 mg/5–10 mL cell culture (Schneemann & Marshall, 1998). VLPs or mutant particles can be expressed in a heterologous system based on baculovirus (Schneemann *et al.*, 1993). (Heterologous expression systems and *in vitro* assembly methods are discussed in Chapter 3.) The heterologous expression combined with the capsid structure has allowed the display of relatively large protein domains on the capsid surface. FHV is relatively new to the field of nanotechnology, and because of its unique facility in protein display, it has primarily been exploited for vaccine development strategies (Manayani *et al.*, 2007) (see Chapter 8).

2.2.8 HK97: A Bacteriophage

HK97 is an *Enterobacteria* phage that infects *E. coli*. The phages consist of a head and a tail, are not enveloped, and encapsidate a linear double-stranded DNA genome. The head is the structure of interest for nanotechnology. The phage head has icosahedral $T = 7$ symmetry and a diameter of 66 nm.

The most intriguing feature of the virions is their maturation process, during which the particles undergo expansion from 54 nm (prohead II) to 66 nm (head II) (Fig. 2.11) (Gertsman *et al.*, 2009; Lee *et al.*, 2008). VNPs in general can be regarded as rigid and robust structures; however, one has to keep in mind that they are also highly dynamic structures. For example, we already mentioned the pH- and metal ion–dependent swelling mechanism of CCMV (see Section 2.2.2). HK97 particles undergo an expansion and increase their size by about 20% during maturation. The matured capsid, called head II, is 66 nm in diameter and extremely thin walled (Gertsman *et al.*, 2009; Lee *et al.*, 2008).

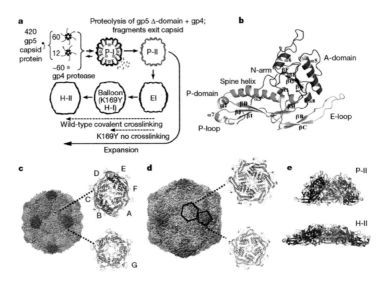

Figure 2.11 HK97 assembly and morphology. (a). The 384-residue gp5 subunit initially assembles into hexameric and pentameric oligomers, termed capsomers, that first assemble to form the prohead I capsid (P-I). The $T = 7$ particle is composed of 12 pentamers and 60 hexamers and encapsidates approximately 60 copies of gp4 protease. Expression with a defective protease produces a prohead I particle that can be disassembled *in vitro* into free capsomers and re-assembled when exposed to specific chemical treatments. When active gp4 is present, particles spontaneously mature to the 13 MDa prohead II (P-II) form after digestion of residues 2–103 from all subunits. Cross-linking occurs in the wild-type particle after formation of the EI state. Cross-links (isopeptide bond) form between Lys 169 and Asn 356 located on different subunits. A cross-link-defective mutant, K169Y, expands to head I, a state nearly identical to balloon minus the cross-links. Wild-type balloon undergoes a final expansion step to head II in which the pentons become more protruded and form one last class of cross-links, with a molecular topology similar to chain mail. (b). Crystal structure of subunit D of prohead II at 3.65 Å. (c) 3.65 Å electron density map (displayed as a solid surface) of the full prohead II capsid, contoured at,1s in Chimera. The prohead II hexamers and pentamers are shown alongside the capsid with the seven subunits of the viral asymmetric subunit labeled A–F for the hexamers and G for the pentamers. (d). A calculated electron density map of the head II capsid shown at 3.65 Å, also rendered at,1s. (e). Prohead II and head II hexamers shown tangential to the capsid surface (rotated 90° from view (c) and (d). Reproduced with permission from Gertsman, I., Gan, L., Guttman, M., Lee, K., Speir, J. A., Duda, R. L., Hendrix, R. W., Komives, E. A., and Johnson, J. E. (2009) An unexpected twist in viral capsid maturation, *Nature*, **458**(7238), 646–650.

HK97 wild-type and mutant particles can be produced and assembled in *E. coli* from the expression of just two viral gene products: gp4 (a protease, which is facilitating maturation) and gp5 (the coat protein). Maturation of the capsids can be triggered *in vitro* by chemical or low-pH treatments (Conway *et al.*, 2001; Gan *et al.*, 2006; Lata *et al.*, 2000; Lee *et al.*, 2008; Wikoff *et al.*, 2000).

It was not until very recently that HK97 became a platform for utilization in nanotechnology. The team led by Johnson (The Scripps Research Institute, La Jolla, CA) is currently exploring the feasibility of the particles for biomedical applications. Its biochemistry and the large interior volume of the matured head make it a potentially interesting candidate for drug-delivery strategies.

2.2.9 P22: A Bacteriophage

The *Salmonella typhimurium* bacteriophage P22 is closely related to HK97 (see Section 2.2.7). Like HK97, the phages consist of a head and a tail, are not enveloped, and encapsidate a linear double-stranded DNA genome. The head has icosahedral $T = 7$ symmetry and a diameter of about 58 nm (Chang *et al.*, 2006). Also for P22, it is the capsid that is of interest for nanotechnology. Wild-type P22 as well as VLPs can be expressed with high yields, and *in vitro* assembly methods have been established (Botstein *et al.*, 1972; Fuller & King, 1982; King *et al.*, 1976; King & Casjens, 1974). A recent pilot study demonstrated that chemical functionalities can be installed on the outer surface of the capsid (Kang *et al.*, 2008), which opens the door for future development of the candidate VNP.

2.2.10 T7: A Bacteriophage

T7 also shares similarities with HK97 and P22. T7 is a double-stranded DNA (linear genome) head-tail coliphage. The head is symmetric with icosahedral $T = 7$ symmetry and about 60 nm in diameter (Agirrezabala *et al.*, 2007).

Empty T7 particles can be produced with ease using various methods: empty capsids are assembled prior to DNA packing and can be isolated (in low yields) at the early stage of the infection (Studier, 1972). T7 ghost particles can also be prepared by osmotic shock with yields of about 55% (Liu *et al.*, 2005), or by alkaline lysis with yields as high as greater than 98% (Liu *et al.*, 2006). The availability of empty particles has led to the exploration of T7 for nanotechnology. Empty T7 phages have been utilized as nanocontainers for the encapsulation or constrained synthesis of materials (Liu *et al.*, 2005, 2006) (discussed in Chapter 5).

2.2.11 MS2: A Bacteriophage

MS2 is an enterobacteriophage from the family *Leviviridae*. The structure of the capsid has been solved to atomic resolution (Golmohammadi *et al.*, 1993); the capsid is self-assembled by 180 identical copies of the coat protein monomer and displays icosahedral $T = 3$ symmetry with a diameter of approximately 27 nm. The genome is a single-stranded, positive-sense RNA genome (Fiers *et al.*, 1976). Native and mutant VNPs can be propagated and isolated from *E. coli* cultures with yields of 30 mg of pure particles per 1 L cell culture (Hooker *et al.*, 2004). Also, VLPs can be efficiently expressed in *E. coli* (Pickett & Peabody, 1993).

MS2 has been studied and utilized for nanotechnology; a range of different chemistries have been applied and developed (see Chapter 4). The main focus on MS2 lies in the development of the particles for biomedical applications such as vaccine strategies and novel therapies (discussed in Chapter 8).

2.2.12 Qβ: A Bacteriophage

The bacteriophages Qβ and MS2 are closely related structurally, but Qβ is more stable due to inter-subunit cross-linking by disuflide bonds (Ashcroft *et al.*, 2005). Like MS2, Qβ is an enterobacteriophage from the family *Leviviridae*. Qβ has symmetric particles of $T = 3$ that are formed by 180 copies of a single coat protein. The genome consists of single-stranded, positive-sense RNA (Golmohammadi *et al.*, 1996).

Qβ VNPs as well as VLPs have been widely used in nanotechnology spanning the fields of materials and medicine and will be featured throughout the chapters of this book.

2.2.13 *Adenovirus*-Based Vectors: A Mammalian Virus

Ads are from the *Adenoviridae* family. There are 51 serotypes of human Ads. Ad serotype 5 is the most commonly studied for applications in biomedical nanotechnology. Ad infections cause illness of the respiratory system and are associated with the common cold. The particles are non-enveloped icosahedral particles with $T = 25$ symmetry and a diameter of 60–90 nm and encapsidate a double-stranded linear DNA genome (Rux & Burnett, 2004; San Martin & Burnett, 2003).

Figure 2.12 shows a representation of the structure of Ads. The particles consist of 12 structural proteins; the main coat proteins are penton, hexon, and fiber, in addition to the hexon- and penton-stabilizing protein pIX. The fiber is responsible for the high-affinity binding of Ad to its receptor, the coxsackie and adenovirus receptor (CAR). The RGD (Arg-Gly-Asp) polypeptide loop that extends the penton is responsible for Ad binding to its secondary cell receptors (integrins $\alpha_v\beta_3$ and $\alpha_v\beta_5$), which triggers its internalization (Rux & Burnett, 2004; San Martin & Burnett, 2003).

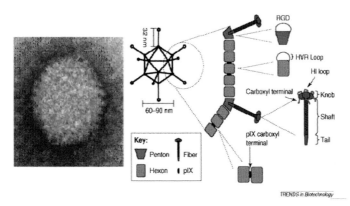

Figure 2.12 Representation of the *Adenovirus* (Ad) structure. Ads have an icosahedral morphology with a core diameter of 60–90 nm and possess a flexible fiber trimer extending 32 nm from each of the 12 vertices. The viral capsid consists of three major proteins: hexon, penton, and fiber, in addition to several proteins, such as pIX, that stabilize hexon and penton. In the left-hand-side picture, a false-colored electron micrograph image of an Ad is shown. Hexon is clearly visible (in green). Sites of penton are identified in blue, and a single fiber is apparent (upper-left corner, in red). The schematic illustration of the viral structure (on the right) shows the relative positions of the capsid proteins for which specific engineering strategies have been developed. The sites that are most suitable for incorporating novel peptide sequences into the capsid structure through genetic engineering are highlighted. These include the hexon hypervariable region (HVR), the penton Arg-Gly-Asp (RGD) loop, the pIX carboxyl terminal, and the fiber knob carboxyl terminal and HI loop. Reproduced with permission from Singh, R., and Kostarelos, K. (2009) Designer *Adenoviruses* for nanomedicine and nanodiagnostics, *Trends Biotechnol.*, **27**(4), 220–229.

Gene therapy. Recombinant Ad vectors along with adeno-associated virus (AAV) vectors (2.2.13) and *Lentivirus*-based vectors[2] have been extensively developed and utilized for therapeutic gene delivery. Gene therapy refers to the correction of a defective gene by insertion of complementary gene functions into cells and tissues to treat a disease. A normal gene may be delivered to replace a non-functional gene and is commonly used to treat a hereditary disease. Ad-based vectors have been the platform of choice for many years for various reasons: Ad vectors are highly efficient for transfecting cells (the

---

[2]*Lentivirus* is also a mammalian virus; it is member of the *Retroviridae* family. As such, its characteristics are transfer and integration of large DNA segments into the host genome. Together with Ads, it is one of the most efficient gene-delivery vectors.

introduction of genes into cells using viral vectors or other nanoparticle platforms is termed transfection), have large capacity for incorporation of DNA cassettes, and have low potential for oncogenesis (the process by which a normal cell is transformed into the cancerous cell) because they do not insert their genome into the host genome (Russell, 2000). Nanotechnology has brought the development of Ad vectors for medical applications to the next level. Novel strategies have been applied that allow to further develop and fine-tune Ad vectors to achieve improved gene delivery. Moreover, new trends have been developed including the development of specifically targeted formulations. Further methods to exploit Ads for imaging purposes and delivery of small molecules are also explored (reviewed in Singh & Kostarelos, 2009) (see also Chapter 8).

2.2.14 Adeno-Associated Virus: A Mammalian Virus

AAV are from the *Parvoviridae* family. AAV infects humans and other primates. Twelve human serotypes are known (AAV-1 to AAV-12). The most popular and extensively utilized serotype is AAV-2. AAV is not known to cause a disease. Productive AAV infection can only occur in the presence of a helper virus, such as Ad (serotype 5) or *Herpes simplex virus* serotype 1 (Handa & Carter, 1979).

The particles are non-enveloped icosahedral particles with $T = 1$ symmetry and a diameter of about 27 nm; the capsids are formed by 60 identical copies of one coat protein. The genome is a single-stranded linear DNA genome; sense and antisense strands are packed into the AAV capsids with equal frequency (reviewed in Daya & Berns, 2008)

The main applications of AAV are in the field of biomedical nanotechnology. AAV-based vectors are utilized in gene therapy because of their propensity to integrate into the human genome in a specific and controlled fashion (discussed in Chapter 8).

2.2.15 *Canine Parvovirus*: A Mammalian Virus

CPV is a mammalian virus from the *Parvoviridae* family that infects canids (dogs). The particles have $T = 1$ symmetry with an average diameter of 26 nm. The capsid is formed by 60 copies of the main coat protein VP2 and a few copies of VP1 and VP3. The genome is linear single-stranded DNA (Tsao *et al.*, 1991). CPV VLPs can be produced for safe use in nanotechnology in mammalian or insect cells by expressing only the VP2 subunit (Yuan & Parrish, 2001).

CPV naturally infects canine cells using the transferrin receptor (Parker *et al.*, 2001). Transferrin is a circulatory iron carrier protein that is in great demand, particularly during cellular growth and proliferation; transferrin receptors are upregulated on cancer cells. The natural receptor specificity of CPV has been exploited for biomedical nanotechnology (discussed in Chapter 8). A pilot study demonstrated that CPV particles are naturally targeted to and internalized by various mammalian tumor cells *in vitro* (Singh *et al.*, 2006).

2.3 ROD-SHAPED VNPs

2.3.1 *Potato Virus X*: A Plant Virus

PVX is the type member of the potexvirus group (Koenig & Lesemann, 1989). The particles are flexible rods with dimensions of 515 nm by 13 nm, and consist of 1270 identical coat protein subunits (Kendall *et al.*, 2008). PVX can be found worldwide in areas where potatoes are grown. Natural hosts are members of the family *Solanaceae*, and *N. benthamiana* (tobacco) plants are also susceptible to mechanical inoculation to produce particles in gram quantities from 1 kg infected leaf material. Infectious cDNA clones of PVX genomic RNA are available, and genetic modifications protocols have been established (Baulcombe *et al.*, 1995). PVX has been used for epitope presentation strategies for vaccine development (Uhde *et al.*, 2005; Uhde-Holzem *et al.*, 2007) and recently also received attention as a VNP platform for materials science (Carette *et al.*, 2007) and medicine (Steinmetz *et al.*, 2009) (see Chapters 4, 7, and 8).

2.3.2 *Tobacco Mosaic Virus*: A Plant Virus

TMV is a rod-shaped tobamovirus. TMV is distributed worldwide and has a broad host range. TMV is transmitted by mechanical contact between plants and man. The particles can be produced in high titers in *Nicotiana tabacum* or *N. benthamiana* plants; yields of up to 2 g/kg infected leaf material can be obtained.

TMV particles contain a single-stranded, positive-sense RNA genome. The capsid is a straight, rigid, tubular structure with dimensions of 18 by 300 nm. It is composed of approximately 2130 identical protein subunits closely packed in a helix with a pitch of 2.3 nm and 16 1/3 subunits per turn (Fig. 2.7). The X-ray structure of the particles has been determined (Namba & Stubbs, 1986). The self-assembly of the TMV rod has been extensively studied (Culver, 2002). *In vitro* assembly and disassembly methods have been established and are exploited for various nanotechnological applications.

TMV particles have been extensively studied for manifold potential applications spanning the fields of medicine and materials. The potential and advances of TMV in the field of nanotechnology will be discussed throughout all the chapters of this book.

2.3.3 M13: A Bacteriophage

The filamentous coliphage M13 is from the family *Inoviridae*. M13 particles are up to 1 μm in length and about 5–6 nm in diameter. The capsid encapsulates a circular single-stranded DNA molecule and is mainly composed of 2700 copies of the major coat protein pVIII. Minor coat protein subunits are found on the ends of the VNP: on one end five copies each of pVII and pIX, and on the other end five copies each of pIII and pVI (Kehoe & Kay, 2005). M13 particles and mutant particles can readily be produced in *E. coli*.

Phage display technology. Besides its manifold applications in materials, M13 has long been exploited as a platform for phage display technologies. Phage display is a high-throughput screening technique that is used to identify peptides that are specifically interacting with molecular proteins or receptors (Arap *et al.*, 2002; Hajitou *et al.*, 2006; Nanda & St. Croix, 2004; Ruoslahti, 2002) or even synthetic materials (for a review see Flynn *et al.*, 2003). In brief, DNA sequences encoding for random peptide sequences are ligated into the pIII or pVIII gene. The peptides are thus displayed on the surface of the phages. The phage is then produced in *E. coli*. The target (protein or inorganic material) is immobilized on a surface and phages are applied. The unbound phages are rinsed off; those that are bound to the target are then eluted and amplified in the expression system. This cycle of binding, elution, and amplification is called biopanning and repeated several times to select the phages with highest selectivity and specificity. The amino acid sequence of the desired peptide is readily obtained by sequencing of the encoding nucleic acid of the selected phage.

Phage display has revolutionized several technologies. The identification of ligands specifically for vasculature and tumor-specific markers has opened the door for the development of targeted therapies (Arap *et al.*, 2002; Hajitou *et al.*, 2006; Nanda & St. Croix, 2004; Ruoslahti, 2002). On the materials side of applications, many peptides that bind selectively to inorganic materials such as metals have been discovered. These peptides allow to bridge biology with materials (for a review see Flynn *et al.*, 2003) and have led to the development of hybrid M13-based battery electrodes (Lee *et al.*, 2009; Nam *et al.*, 2006, 2008). M13 particles have been extensively studied and developed for application in materials and will be discussed in great detail in Chapters 6 and 7.

2.3.4 *Sulfolobus islandicus* Rod-Shaped Virus 2: An Archaeal Virus

SIRV2 is an archaeal virus from the *Rudiviridae* family, which is a group of non-enveloped, rod-shaped viruses with linear double-stranded DNA genomes (Prangishvili *et al.*, 2006). SIRV2 infects strains of the gases and steam), hot, acidic spring in Iceland with temperature 88°C and pH 2.5. SIRV2 grows in a truly extreme environment. Due to its hyperthermophilic archaeal genus

A

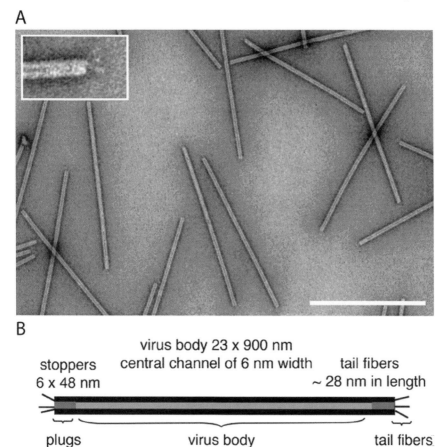

B

virus body 23 x 900 nm
central channel of 6 nm width

stoppers
6 x 48 nm

tail fibers
~ 28 nm in length

plugs   virus body   tail fibers

Figure 2.13 Structure of SIRV2 particles. (A). TEM of negatively stained SIRV2 particles. Inset shows a high-resolution image of the end structure of the capsid. The scale bar is 500 nm. (B). Schematic depiction of SIRV2 particles. Reproduced with permission from Steinmetz, N. F., Bize, A., Findlay, K. C., Lomonossoff, G. P., Manchester, M., Evans, D. J., and Prangishvili, D. (2008) Site-specific and spatially controlled addressability of a new viral nanobuilding block: *Sulfolobus islandicus* rod-shaped virus 2, *Adv. Funct. Mater.*, **18**, 3478–3486.

Sulfolobus. These organisms were isolated from a solfataric (a volcanic area that gives off sulfurous hyperthermostable and acid-resistant nature, SIRV2 is potentially an extremely stable VNP, and is therefore a highly interesting candidate for nanotechnology. Studies toward utilizing SIRV2 as a VNP platform are still in their infancy, but are promising (Steinmetz *et al.*, 2008).

The rod-shaped SIRV2 particles have dimensions of 23 nm by 900 nm with a central channel of about 6 nm. Multiple copies of the major coat protein form the virion body with a periodicity of 4.3 nm. Three tail fibers of about 28 nm in length are found at either end of the VNPs, and it is indicated that the end structures are formed by a high-molecular-weight minor coat protein (Prangishvili *et al.*, 1999) (Fig. 2.13). It was shown that the virus body and the tail fibers could be selectively labeled using different chemistries (Steinmetz *et al.*, 2008) (more details are given in Chapter 4). The ability to install labels and functionalities on the virion with spatial control in combination with its thermo- and pH-stability makes SIRV2 an interesting candidate for nanotechnology.

2.4 ENVELOPED VNPs

2.4.1 *Chilo Iridescent Virus*: An Invertebrate Virus

CIV is an invertebrate virus from the family *Iridoviridae* that mainly infects insects. The capsids have icosahedral symmetry and are enveloped. CIV has a linear double-stranded DNA genome (from ICTV database). It is the largest VNP that has been utilized for nanotechnology, although there has only been one report in which the capsids (without the envelope) were exploited as templates for metallization (Radloff *et al.*, 2005) (discussed in Chapter 6). The particles are 185 nm in diameter and have $T = 147$ symmetry (Yan *et al.*, 2000).

2.4.2 *Rubella Virus*: A Mammalian Virus

RV is from the family *Togaviridae*. RV is the infectious agent causing the disease Rubella, which is commonly known as German measles (because it has been first described by the German physician Friedrich Hoffman in the mid-18th century). The name rubella is derived from Latin and means "little red."

The virions have icosahedral $T = 4$ symmetry and are enveloped with a measure of 60–70 nm in diameter. The genome is a single molecule of linear, positive-sense, single-stranded RNA (from the ICTV; http://www.ncbi.nlm.nih.gov/ICTVdb).

A few studies have been published in which enveloped VNPs were used for a nanotechnology application. For safe use in materials VLPs or UV-inactivated particles were exploited and integrated into polyelectrolyte thin films (Fischlechner *et al.*, 2005, 2006, 2007; Toellner *et al.*, 2006). The incorporation of VNPs into supported membranes may lead to the development of novel nanoscale devices for diagnostics. The immobilization of VNPs on supports and potential applications is discussed in Chapter 7.

2.5 SUMMARY: THE MANIFOLD VNP BUILDING BLOCKS

To date, approximately 20 of 5,000 discovered viruses have been exploited for applications in nanotechnology. Each VNP has different characteristics that can be exploited. Some platforms have been extensively studied; others are still new to the field. Based on the enormous variety of structural and physicochemical properties available, viral nanotechnology has great potential and it is expected that over the years more and more platforms will be evaluated.

References

Agirrezabala, X., Velazquez-Muriel, J. A., Gomez-Puertas, P., Scheres, S. H., Carazo, J. M., and Carrascosa, J. L. (2007) Quasi-atomic model of bacteriophage T7 procapsid shell: insights into the structure and evolution of a basic fold, *Structure*, **15**(4), 461–472.

Ahlquist, P., Dasgupta, R., and Kaesberg, P. (1984) Nucleotide sequence of the brome mosaic virus genome and its implications for viral replication, *J. Mol. Biol.*, **172**(4), 369–383.

Ahlquist, P., Luckow, V., and Kaesberg, P. (1981) Complete nucleotide sequence of brome mosaic virus RNA3, *J. Mol. Biol.*, **153**(1), 23–38.

Arap, W., Haedicke, W., Bernasconi, M., Kain, R., Rajotte, D., Krajewski, S., Ellerby, H. M., Bredesen, D. E., Pasqualini, R., and Ruoslahti, E. (2002) Targeting the prostate for destruction through a vascular address, *Proc. Natl. Acad. Sci. USA*, **99**(3), 1527–1531.

Ashcroft, A. E., Lago, H., Macedo, J. M., Horn, W. T., Stonehouse, N. J., and Stockley, P. G. (2005) Engineering thermal stability in RNA phage capsids via disulphide bonds, *J. Nanosci. Nanotechnol.*, **5**(12), 2034–2041.

Bancroft, J. B. (1962) Purification and properties of bean pod mottle virus and associated centrifugal and electrophoretic components, *Virology*, **16**, 419–427.

Bancroft, J. B., Bracker, C. E., and Wagner, G. W. (1969) Structures derived from cowpea chlorotic mottle and brome mosaic virus protein, *Virology*, **38**(2), 324–335.

Basnayake, V. R., Sit, T. L., and Lommel, S. A. (2006) The genomic RNA packaging scheme of *Red clover necrotic mosaic virus*, *Virology*, **345**(2), 532–539.

Baulcombe, D. C., Chapman, S., Santa, and Cruz, S. (1995) Jellyfish green fluorescent protein as a reporter for virus infections, *Plant J.*, **7**(6), 1045–1053.

Botstein, D., Chan, R. K., and Waddell, C. H. (1972) Genetics of bacteriophage P22. II. Gene order and gene function, *Virology*, **49**(1), 268–282.

Bruening, G., and Agrawal, H. O. (1967) Infectivity of a mixture of cowpea mosaic virus ribonucleoprotein components, *Virology*, **32**(2), 306–320.

Brumfield, S., Willits, D., Tang, L., Johnson, J. E., Douglas, T., and Young, M. (2004) Heterologous expression of the modified coat protein of *Cowpea chlorotic mottle bromovirus* results in the assembly of protein cages with altered architectures and function, *J. Gen. Virol.*, **85**(Pt 4), 1049–1053.

Carette, N., Engelkamp, H., Akpa, E., Pierre, S. J., Cameron, N. R., Christianen, P. C., Maan, J. C., Thies, J. C., Weberskirch, R., Rowan, A. E., Nolte, R. J., Michon, T., and Van Hest, J. C. (2007) A virus-based biocatalyst, *Nat. Nanotechnol.*, **2**(4), 226–229.

Caspar, D. L., and Klug, A. (1962) Physical principles in the construction of regular viruses, *Cold Spring Harb. Symp. Quant. Biol.*, **27**, 1–24.

Chang, J., Weigele, P., King, J., Chiu, W., and Jiang, W. (2006) Cryo-EM asymmetric reconstruction of bacteriophage P22 reveals organization of its DNA packaging and infecting machinery, *Structure*, **14**(6), 1073–1082.

Chen, C., Daniel, M. C., Quinkert, Z. T., De, M., Stein, B., Bowman, V.D., Chipman, P. R., Rotello, V. M., Kao, C. C., and Dragnea, B. (2006) Nanoparticle-templated assembly of viral protein cages, *Nano. Lett.*, **6**(4), 611–615.

Conway, J. F., Wikoff, W. R., Cheng, N., Duda, R. L., Hendrix, R. W., Johnson, J. E., and Steven, A. C. (2001) Virus maturation involving large subunit rotations and local refolding, *Science*, **292**(5517), 744–748.

Culver, J. N. (2002) Tobacco mosaic virus assembly and disassembly: determinants in pathogenicity and resistance, *Annu. Rev. Phytopathol.*, **40**, 287–308.

Dixit, S. K., Goicochea, N. L., Daniel, M. C., Murali, A., Bronstein, L., De, M., Stein, B., Rotello, V. M., Kao, C. C., and Dragnea, B. (2006) Quantum dot encapsulation in viral capsids, *Nano Lett.*, **6**(9), 1993–1999.

Dragnea, B., Chen, C., Kwak, E. S., Stein, B., and Kao, C. C. (2003) Gold nanoparticles as spectroscopic enhancers for *in vitro* studies on single viruses, *J. Am. Chem. Soc.*, **125**(21), 6374–6375.

Fiers, W., Contreras, R., Duerinck, F., Haegeman, G., Iserentant, D., Merregaert, J., Min Jou, W., Molemans, F., Raeymaekers, A., Van den Berghe, A., Volckaert, G., and Ysebaert,

M. (1976) Complete nucleotide sequence of bacteriophage MS2 RNA: primary and secondary structure of the replicase gene, *Nature*, **260**(5551), 500–507.

Fischlechner, M., Reibetanz, U., Zaulig, M., Enderlein, D., Romanova, J., Leporatti, S., Moya, S., and Donath, E. (2007) Fusion of enveloped virus nanoparticles with polyelectrolyte-supported lipid membranes for the design of bio/nonbio interfaces, *Nano. Lett.*, **7**, 3540–3546.

Fischlechner, M., Toellner, L., Messner, P., Grabherr, R., and Donath, E. (2006) Virus-engineered colloidal particles: a surface display system, *Angew. Chem. Int. Ed. Engl.*, **45**(5), 784–789.

Fischlechner, M., Zschornig, O., Hofmann, J., and Donath, E. (2005) Engineering virus functionalities on colloidal polyelectrolyte lipid composites, *Angew. Chem. Int. Ed. Engl.*, **44**(19), 2892–2895.

Fisher, A. J., McKinney, B. R., Schneemann, A., Rueckert, R. R., and Johnson, J. E. (1993) Crystallization of viruslike particles assembled from flock house virus coat protein expressed in a baculovirus system, *J. Virol.*, **67**(5), 2950–2953.

Flynn, C. E., Lee, S.-W., Peelle, B. R., and Belcher, A. M. (2003) Viruses as vehicles for growth, organisation and assembly of materials, *Acta. Materialia.*, **51**, 5867–5880.

Frey, T. K. (1994) Molecular biology of rubella virus, *Adv. Virus. Res.*, **44**, 69–160.

Fuller, M. T., and King, J. (1982) Assembly *in vitro* of bacteriophage P22 procapsids from purified coat and scaffolding subunits, *J. Mol. Biol.*, **156**(3), 633–665.

Gan, L., Speir, J. A., Conway, J. F., Lander, G., Cheng, N., Firek, B. A., Hendrix, R. W., Duda, R. L., Liljas, L., and Johnson, J. E. (2006) Capsid conformational sampling in HK97 maturation visualized by X-ray crystallography and cryo-EM, *Structure*, **14**(11), 1655–1665.

Gertsman, I., Gan, L., Guttman, M., Lee, K., Speir, J. A., Duda, R. L., Hendrix, R. W., Komives, E. A., and Johnson, J. E. (2009) An unexpected twist in viral capsid maturation, *Nature*, **458**(7238), 646–650.

Golmohammadi, R., Fridborg, K., Bundule, M., Valegard, K., and Liljas, L. (1996) The crystal structure of bacteriophage Q beta at 3.5 A resolution, *Structure*, **4**(5), 543–554.

Golmohammadi, R., Valegard, K., Fridborg, K., and Liljas, L. (1993) The refined structure of bacteriophage MS2 at 2.8 A resolution, *J. Mol. Biol.*, **234**(3), 620–639.

Hajitou, A., Pasqualini, R., and Arap, W. (2006) Vascular targeting: recent advances and therapeutic perspectives, *Trends Cardiovasc. Med.*, **16**(3), 80–88.

Hooker, J. M., Kovacs, E. W., and Francis, M. B. (2004) Interior surface modification of bacteriophage MS2, *J. Am. Chem. Soc.*, **126**(12), 3718–3719.

Johnson, J. E., and Speir, J. A. (1997) Quasi-equivalent viruses: a paradigm for protein assemblies, *J. Mol. Biol.*, **269**(5), 665–675.

Kang, S., Lander, G. C., Johnson, J. E., and Prevelige, P. E. (2008) Development of bacteriophage P22 as a platform for molecular display: genetic and chemical modifications of the procapsid exterior surface, *ChemBioChem,* **9**(4), 514–518.

Kehoe, J. W., and Kay, B. K. (2005) Filamentous phage display in the new millennium, *Chem. Rev.,* **105**(11), 4056–4072.

Kendall, A., McDonald, M., Bian, W., Bowles, T., Baumgarten, S. C., Shi, J., Stewart, P. L., Bullitt, E., Gore, D., Irving, T. C., Havens, W. M., Ghabrial, S. A., Wall, J. S., and Stubbs, G. (2008) Structure of flexible filamentous plant viruses, *J. Virol.,* **82**(19), 9546–9554.

King, J., Botstein, D., Casjens, S., Earnshaw, W., Harrison, S., and Lenk, E. (1976) Structure and assembly of the capsid of bacteriophage P22, *Philos. Trans. R. Soc. Lond. B. Biol. Sci.,* **276**(943), 37–49.

King, J., and Casjens, S. (1974) Catalytic head assembling protein in virus morphogenesis, *Nature,* **251**(5471), 112–119.

Knipe, D. M., and Howley P. M. (2001) *Fundamental Virology*, 4th edn, Philadelphia, Lippincott Williams & Wilkins.

Koenig, R., and Lesemann, D. E. (1989) *Potato virus X, potexvirus group*, vol. 354, Warwick, Association of Applied Biologists.

Koshiyama, T., Yokoi, N., Ueno, T., Kanamaru, S., Nagano, S., Shiro, Y., Arisaka, F., and Watanabe, Y. (2008) Molecular design of heteroprotein assemblies providing a bionanocup as a chemical reactor, *Small,* **4**(1), 50–54.

Krishna, N. K., and Schneemann, A. (1999) Formation of an RNA heterodimer upon heating of nodavirus particles, *J. Virol.,* **73**(2), 1699–1703.

Lata, R., Conway, J. F., Cheng, N., Duda, R. L., Hendrix, R. W., Wikoff, W. R., Johnson, J. E., Tsuruta, H., and Steven, A. C. (2000) Maturation dynamics of a viral capsid: visualization of transitional intermediate states, *Cell,* **100**(2), 253–263.

Lee, K. K., Gan, L., Tsuruta, H., Moyer, C., Conway, J. F., Duda, R. L., Hendrix, R. W., Steven, A. C., and Johnson, J. E. (2008) Virus capsid expansion driven by the capture of mobile surface loops, *Structure,* **16**(10), 1491–1502.

Lee, Y. J., Yi, H., Kim, W. J., Kang, K., Yun, D. S., Strano, M. S., Ceder, G., and Belcher, A. M. (2009) Fabricating genetically engineered high-power lithium-ion batteries using multiple virus genes, *Science,* **324**(5930), 1051–1055.

Liepold, L. O., Revis, J., Allen, M., Oltrogge, L., Young, M., and Douglas, T. (2005) Structural transitions in *Cowpea chlorotic mottle virus* (CCMV), *Phys. Biol.,* **2**(4), S166–S172.

Lin, T., Chen, Z., Usha, R., Stauffacher, C. V., Dai, J. B., Schmidt, T., and Johnson, J. E. (1999) The refined crystal structure of cowpea mosaic virus at 2.8 A resolution, *Virology,* **265**(1), 20–34.

Liu, C. M., Chung, S.-H., Jin, Q., Sutton, A., Yan, F., Hoffmann, A., Kay, B. K., Bader, S. D., Makowski, L., and Chen, L. (2006) Magnetic viruses via nano-capsid templates, *J. Magn. Magn. Mater.,* **302**, 47–51.

Liu, C. M., Jin, Q., Sutton, A., and Chen, L. (2005) A novel fluorescent probe: europium complex hybridized T7 phage, *Bioconjug. Chem.*, **16**(5), 1054–1057.

Lomonossoff, G. P., and Johnson, J. E. (1991) The synthesis and structure of comovirus capsids, *Prog. Biophys. Mol. Biol.*, **55**(2), 107–137.

Lomonossoff, G. P., and Shanks, M. (1983) The nucleotide sequence of cowpea mosaic virus B RNA, *EMBO J.*, **2**(12), 2253–2285.

Lomonossoff, G. P., and Shanks, M. (1999) *Comoviruses (Comoviridae)*, in *Encyclopaedia of Virology*, 2nd edn, Academic Press.

Loo, L., Guenther, R. H., Basnayake, V. R., Lommel, S. A., and Franzen, S. (2006) Controlled encapsidation of gold nanoparticles by a viral protein shell, *J. Am. Chem. Soc.*, **128**(14), 4502–4503.

Loo, L., Guenther, R. H., Lommel, S. A., and Franzen, S. (2007) Encapsidation of nanoparticles by red clover necrotic mosaic virus, *J. Am. Chem. Soc.*, **129**(36), 11111–11117.

Lucas, R. W, Larson, S. B, and McPherson, A. (2002) The crystallographic structure of brome mosaic virus, *J. Mol. Biol.*, **317**(1), 95–108.

Manayani, D. J., Thomas, D., Dryden, K. A., Reddy, V., Siladi, M. E., Marlett, J. M., Rainey, G. J., Pique, M. E., Scobie, H. M., Yeager, M., Young, J. A., Manchester, M., and Schneemann, A. (2007) A viral nanoparticle with dual function as an anthrax antitoxin and vaccine, *PLoS Pathog.*, **3**(10), 1422–1431.

Mao, C., Solis, D. J., Reiss, B. D., Kottmann, S. T., Sweeney, R. Y., Hayhurst, A., Georgiou, G., Iverson, B., and Belcher, A. M. (2004) Virus-based toolkit for the directed synthesis of magnetic and semiconducting nanowires, *Science*, **303**(5655), 213–217.

Nam, K. T., Kim, D. W., Yoo, P. J., Chiang, C. Y., Meethong, N., Hammond, P. T., Chiang, Y. M., and Belcher, A. M. (2006) Virus-enabled synthesis and assembly of nanowires for lithium ion battery electrodes, *Science*, **312**(5775), 885–888.

Nam, K. T., Peelle, B. R., Lee, S. W., and Belcher, A. M. (2004) Genetically driven assembly of nanorings based on the M13 virus, *Nano Lett.*, **4**(1), 23–27.

Nam, K. T., Wartena, R., Yoo, P. J., Liau, F. W., Lee, Y. J., Chiang, Y. M., Hammond, P. T., and Belcher, A. M. (2008) Stamped microbattery electrodes based on self-assembled M13 viruses, *Proc. Natl. Acad. Sci. USA*, **105**(45), 17227–17231.

Namba, K., and Stubbs, G. (1986) Structure of tobacco mosaic virus at 3.6 A resolution: implications for assembly, *Science*, **231**(4744), 1401–1406.

Nanda, A., and St. Croix, B. (2004) Tumor endothelial markers: new targets for cancer therapy, *Curr. Opin. Oncol.*, **16**, 44–49.

Parker, J. S., Murphy, W. J., Wang, D., O'Brien, S. J., and Parrish, C. R. (2001) Canine and feline parvoviruses can use human or feline transferrin receptors to bind, enter, and infect cells, *J. Virol.*, **75**(8), 3896–3902.

Pfeiffer, P., Herzog, M., and Hirth, L. (1976) RNA viruses: stabilization of brome mosaic virus, *Philos. Trans. R. Soc. Lond. B. Biol. Sci.*, **276**(943), 99–107.

Pfeiffer, P., and Hirth, L. (1974) Aggregation states of brome mosaic virus protein, *Virology*, **61**(1), 160–167.

Pickett, G. G., and Peabody, D. S. (1993) Encapsidation of heterologous RNAs by bacteriophage MS2 coat protein, *Nucleic Acids Res.*, **21**(19), 4621–4626.

Prangishvili, D., Arnold, H. P., Gotz, D., Ziese, U., Holz, I., Kristjansson, J. K., and Zillig, W. (1999) A novel virus family, the Rudiviridae: structure, virus-host interactions and genome variability of the sulfolobus viruses SIRV1 and SIRV2, *Genetics*, **152**(4), 1387–1396.

Prangishvili, D., Forterre, P., and Garrett, R. A. (2006) Viruses of the Archaea: a unifying view, *Nat. Rev. Microbiol.*, **4**(11), 837–848.

Radloff, C., Vaia, R. A., Brunton, J., Bouwer, G. T., and Ward, V. K. (2005) Metal nanoshell assembly on a virus bioscaffold, *Nano Lett.* **5**(6), 1187–1191.

Ren, Y., Wong, S. M., and Lim, L. Y. (2006) *In vitro*-reassembled plant virus-like particles for loading of polyacids, *J. Gen. Virol.*, **87**(Pt 9), 2749–2754.

Ren, Y., Wong, S. M., and Lim, L. Y. (2007) Folic acid-conjugated protein cages of a plant virus: a novel delivery platform for doxorubicin, *Bioconjug. Chem.*, **18**(3), 836–843.

Rossmann, M. G., and Johnson, J. E. (1989) Icosahedral RNA virus structure, *Annu. Rev. Biochem.*, **58**, 533–573.

Roy, P. (2005) *Advances in Virus Research*, Elsevier Academic Press.

Ruoslahti, E. (2002) Specialization of tumor vasculature, *Nat. Rev. Cancer*, **2**, 83–90.

Russell, W. C. (2000) Update on adenovirus and its vectors, *J. Gen. Virol.*, **81**(Pt 11), 2573–2604.

Rux, J. J., and Burnett, R. M. (2004) Adenovirus structure, *Hum. Gene Ther.*, **15**(12), 1167–1176.

San Martin, C., and Burnett, R. M. (2003) Structural studies on adenoviruses, *Curr. Top Microbiol. Immunol.*, **272**, 57–94.

Saunders, K., Sainsbury, F., and Lomonossoff, G. P. (2009) Efficient generation of cowpea mosaic virus empty virus-like particles by the proteolytic processing of precursors in insect cells and plants, *Virology*, 393(2):329–337.

Schlick, T. L., Ding, Z., Kovacs, E. W., and Francis, M. B. (2005) Dual-surface modification of the tobacco mosaic virus, *J. Am. Chem. Soc.*, **127**, 3718–3723.

Schneemann, A., Dasgupta, R., Johnson, J. E, and Rueckert, R. R. (1993) Use of recombinant baculoviruses in synthesis of morphologically distinct viruslike particles of flock house virus, a nodavirus, *J. Virol.*, **67**(5), 2756–2763.

Schneemann, A., and Marshall, D. (1998) Specific encapsidation of nodavirus RNAs is mediated through the C terminus of capsid precursor protein alpha, *J. Virol.*, **72**(11), 8738–8746.

Schneemann, A., and Young, M. J. (2003) Viral assembly using heterologous expression systems and cell extracts, *Adv. Protein Chem.*, **64,** 1–36.

Shanks, M., and Lomonossoff, G. P. (2000) Co-expression of the capsid proteins of *Cowpea mosaic virus* in insect cells leads to the formation of virus-like particles, *J. Gen. Virol.*, **81**(Pt 12), 3093–3097.

Shenton, W., Douglas, T., Young, M., Stubbs, G., and Mann, S. (1999) Inorganic-organic nanotube composites from template mineralization of tobacco mosaic virus, *Adv. Mater.*, **11,** 253–156.

Sherman, M. B., Guenther, R. H., Tama, F., Sit, T. L., Brooks, C. L., Mikhailov, A. M., Orlova, E. V., Baker, T. S., and Lommel, S. A. (2006) Removal of divalent cations induces structural transitions in *Red clover necrotic mosaic virus*, revealing a potential mechanism for RNA release, *J. Virol.*, **80,** 10395–10406.

Singh, P., Destito, G., Schneemann, A., and Manchester, M. (2006) Canine parvovirus-like particles, a novel nanomaterial for tumor targeting, *J. Nanobiotechnol.*, **4,** 2.

Singh, R., and Kostarelos, K. (2009) Designer adenoviruses for nanomedicine and nanodiagnostics, *Trends Biotechnol.*, **27**(4), 220–229.

Sit, T. L., Vaewhongs, A. A., and Lommel, S. A. (1998) RNA-mediated trans-activation of transcription from a viral RNA, *Science*, **281**(5378), 829–832.

Speir, J. A., Munshi, S., Wang, G., Baker, T. S., and Johnson, J. E. (1995) Structures of the native and swollen forms of cowpea chlorotic mottle virus determined by X-ray crystallography and cryo-electron microscopy, *Structure*, **3**(1), 63–78.

Steinmetz, N. F., Bize, A., Findlay, K. C., Lomonossoff, G. P., Manchester, M., Evans, D. J., and Prangishvili, D. (2008) Site-specific and spatially controlled addressability of a new viral nanobuilding block: *Sulfolobus islandicus* rod-shaped virus 2, *Adv. Funct. Mater.*, **18,** 3478–3486.

Steinmetz, N. F., Mertens, M. E., Taurog, R. E., Johnson, J. E., Commandeur, U., Fischer, R., and Manchester, M. (2009) Potato virus X as a novel platform for potential biomedical applications, *Nano Lett., 10(1), 305–312.*

Studier, F. W. (1972) Bacteriophage T7, *Science*, **176**(33), 367–376.

Sun, J., DuFort, C., Daniel, M. C., Murali, A., Chen, C., Gopinath, K., Stein, B., De, M., Rotello, V. M., Holzenburg, A., Kao, C.C., and Dragnea, B. (2007) Core-controlled polymorphism in virus-like particles, *Proc. Natl. Acad Sci. USA,* **104**(4), 1354–1359.

Toellner, L., Fischlechner, M., Ferko, B., Grabherr, R. M., and Donath, E. (2006) Virus-coated layer-by-layer colloids as a multiplex suspension array for the detection and quantification of virus-specific antibodies, *Clin. Chem.*, **52**(8), 1575–1583.

Tsao, J., Chapman, M. S., Agbandje, M., Keller, W., Smith, K., Wu, H., Luo, M., Smith, T. J., Rossmann, M. G., Compans, R. W., *et al.* (1991) The three-dimensional structure of canine parvovirus and its functional implications, *Science*, **251**(5000), 1456–1464.

Tseng, R. J., Tsai, C., Ma, L., Ouyang, J., Ozkan, C. S., Yang, Y. (2006) Digital memory device based on tobacco mosaic virus conjugated with nanoparticles, *Nat. Nanotechnol.*, **1**, 72–77.

Ueno, T., Koshiyama, T., Tsuruga, T., Goto, T., Kanamaru, S., Arisaka, F., and Watanabe, Y. (2006) Bionanotube tetrapod assembly by in situ synthesis of a gold nanocluster with (Gp5-His6)3 from bacteriophage T4, *Angew. Chem. Int. Ed. Engl.*, **45**(27), 4508–4512.

Uhde, K., Fischer, R., and Commandeur, U. (2005) Expression of multiple foreign epitopes presented as synthetic antigens on the surface of *Potato virus X* particles, *Arch. Virol.*, **150**(2), 327–340.

Uhde-Holzem, K., Fischer, R., and Commandeur, U. (2007) Genetic stability of recombinant potato virus X virus vectors presenting foreign epitopes, *Arch. Virol.*, **152**(4), 805–811.

van Wezenbeek, P., Verver, J., Harmsen, J., Vos, P., and van Kammen, A. (1983) Primary structure and gene organization of the middle-component RNA of cowpea mosaic virus, *EMBO. J.*, **2**(6), 941–946.

Wikoff, W. R., Liljas, L., Duda, R. L., Tsuruta, H., Hendrix, R. W., and Johnson, J. E. (2000) Topologically linked protein rings in the bacteriophage HK97 capsid, *Science*, **289**(5487), 2129–2133.

Wu, G. J., and Bruening, G. (1971) Two proteins from cowpea mosaic virus, *Virology*, **46**(3), 596–612.

Yan, X., Olson, N. H., Van Etten, J. L., Bergoin, M., Rossmann, M. G., and Baker, T. S. (2000) Structure and assembly of large lipid-containing dsDNA viruses, *Nat. Struct. Biol.*, **7**(2), 101–103.

Yi, H., Rubloff, G. W., and Culver, J. N. (2007) TMV microarrays: hybridization-based assembly of DNA-programmed viral nanotemplates, *Langmuir*, **23**(5), 2663–2667.

Yoo, P. J., Nam, K. T., Qi, J. F., Lee, S. K., Park, J., Belcher, A. M., and Hammond, P. T. (2006) Spontaneous assembly of viruses on multilayered polymer surfaces, *Nat. Mater.*, **5**(3), 234–240.

Yuan, W., and Parrish, C. R. (2001) Canine parvovirus capsid assembly and differences in mammalian and insect cells, *Virology*, **279**(2), 546–557.

Zhao, X., Fox, J. M., Olson, N. H., Baker, T. S., and Young, M. J. (1995) *In vitro* assembly of cowpea chlorotic mottle virus from coat protein expressed in *Escherichia coli* and *in vitro*-transcribed viral cDNA, *Virology*, **207**(2), 486–494.

Chapter 3

PRODUCTION OF VNPs, VLPs, AND CHIMERAS

Many viruses accumulate to high titers in their natural hosts; plant viruses, for example, can be isolated in gram scales from a kilogram of infected leaf material. Bacteriophages also accumulate to high titers (gram bacteriophage per liter cell culture), and production of bacteriophages can be scaled up using modern fermentation techniques. As diverse applications in nanotechnology progress toward industrial practice and clinical trials, there is demand for reliable and large-scale production of viral nanoparticles (VNPs). It is desired that the expression system gives high flexibility, allowing production of various mutant VNPs with altered surface properties such as modification of amino acids, or chimeras (VNPs displaying foreign peptide sequences) with ease, at low cost, and in a time-efficient manner. For safe use in materials and medicine, it is generally desirable that the VNP be replication-deficient. Various methods have been developed that allow extraction of the nucleic acid from assembled particles to render them non-infectious. Alternatively, the virus can be inactivated using physical methods such as UV irradiation. However, these methods introduce additional steps in the manufacturing process and can be cumbersome and inefficient.

Heterologous expression systems provide an attractive alternative for the production of non-infectious particles. Heterologous expression systems are highly efficient, extremely flexible, and allow the production of virus-like particles (VLPs; particles that are devoid of genomic nucleic acid). Phages and mammalian viruses are commonly expressed as VLPs in heterologous systems. Plant viruses are expressed either in a heterologous system or in the greenhouse in their natural hosts. Various production strategies and assembly techniques are discussed in this chapter.

Viral Nanoparticles: *Tools for Materials Science and Biomedicine*
By Nicole F. Steinmetz and Marianne Manchester
Copyright © 2011 by Pan Stanford Publishing Pte. Ltd.
www.panstanford.com

3.1 PRODUCTION OF VNPs IN THE GREENHOUSE

A broad range of plant VNPs has been exploited for nanotechnology and biotechnology. The most extensively studied particles are the icosahedrons *Brome mosaic virus* (BMV), *Cowpea chlorotic mottle virus* (CCMV), *Cowpea mosaic virus* (CPMV), *Red clover necrotic mottle virus* (RCNMV), and the rod-shaped particle *Tobacco mosaic virus* (TMV), all of which can be produced in large scales in their natural plant hosts. Plant viruses typically accumulate to high titers in the plants and can be extracted in yields of 1–2 g/kg infected leaf material. The production species for BMV is barley (*Hordeum vulgare*). CCMV and CPMV are produced in black-eyed pea (*Vigna unguiculata*). RCNMV is produced in the common pea (*Phaseolus vulgaris*), and TMV is produced in tobacco plants (*Nicotiana tabacum*). For further information on plant viruses and their hosts, the reader is referred to the Description of Plant Viruses database (DPV; http://www.dpvweb.net).

Infection of plants can be achieved by mechanical inoculation using either homogenized plant tissue from infected leaves, or purified VNPs. An infection can also be achieved using linearized plasmids containing a complementary DNA (cDNA) copy of the viral genomes or *in vitro* RNA transcripts of the genomes (the RNA transcript serves as a template for protein expression during translation); these methods are typically used when generating mutants or chimeras (Section 3.1.1). To infect about 50 plants, one begins with just a few infected leaves. The plant material is homogenized using either mortar and pestle or a blender; the latter is applicable for large-scale inoculations. The plant material is then suspended in buffer and filtered through cheesecloth (the composition of the extraction buffer depends on the VNP; for many viruses 0.01–0.1 M phosphate buffers are used). The leaf homogenate is then rubbed onto the surface of leaves in such a way to break cells on the surface but without causing too much mechanical damage. Carborundum (silicon carbide, SiC) can be applied onto the leaves to help create lesions in the cell walls on the leaf surface and to facilitate viral cell entry.

After initial inoculation, viruses typically spread within the plant via cell-to-cell and long-distance movement. Long-distance movement is through the vascular tissue along with the photoassimilates and the direction is source-to-sink (Mekuria *et al.*, 2008; Roberts *et al.*, 1997; Silva *et al.*, 2002). "Source" refers to leaves that are photosynthetically active (these are developed leaves), and "sink" refers to the developing tissue (young developing leaves). To facilitate systemic spread primary leaves of young plants are infected, and the infection proceeds as the plant develops. At 2–3 weeks post-inoculation, leaves can be harvested and stored frozen indefinitely until further processing.

Purification of VNPs from infected leaf material is straightforward; protocols and references can be found at the DPV (http://www.dpvweb.net). Infected leaves may be homogenized using mortar and pestle or a blender and the appropriate buffer. The homogenized material is then filtered through cheesecloth. The virus is purified from other plant material using differential ultracentrifugation and ultrapelleting. Some protocols include additional chloroform/butanol extraction and polyethylene glycol precipitation steps. In any case, the purification protocols are simple, straightforward, and fast. Using these methods, one person can easily purify up to 200 mg of virus from 100–200 g infected leaf material (typically obtained from around 50 plants) in a day.

3.1.1 Chimeric Virus Technology Using Plant-Based Expression Systems

One of the advantages of VNPs over any synthetic nanomaterial is the ability to genetically engineer the particles, in addition to modifying and functionalizing the particles using chemistry. Genetic engineering allows one to fine-tune the surface properties of the particles. Amino acids can be deleted, changed, or introduced. Short peptide sequences can be introduced in surface loops; even whole proteins can be incorporated into the viral capsids. Examples of mutant particles, their applications, and some basic principles of chimeric virus technology are discussed in Section 3.7.

Genetic engineering of VNPs refers to the manipulation of the genome, which results in modifications on the protein level. All plant viruses currently in use for nanotechnology applications have RNA genomes. The genomes have been sequenced, and the genetic information is available at the National Center for Biotechnology Information (NCBI) database (http://www.ncbi.nlm.nih.gov). To perform genetic modifications, a cDNA copy of the genome is required. The cDNA is the complementary strand of the genome RNA, which can be synthesized by reverse transcription. The cDNA can then be amplified as double-stranded DNA using PCR techniques and inserted into a cloning or expression vector. At this stage, any standard cloning or mutagenesis procedure can be applied in order to introduce the desired modification. For detailed background information on cloning techniques, the reader is referred to textbooks in the fields of molecular biology and biochemistry.

Once the genetic modification has been introduced into the cDNA clone, mutant particles can be produced using either a heterologous expression system (Section 3.2) or a plant-based expression system. Several methods for the production of infectious mutant VNPs in plants have been developed. Early methods involved *in vitro* transcription of infectious RNA transcripts, which were then used as an inoculum (Ahlquist *et al.*, 1984; Ahlquist & Janda, 1984; Allison *et al.*, 1988; Dessens & Lomonossoff, 1993; Meshi *et al.*, 1986; Xiong & Lommel, 1991). Current methods involve inoculating plants with cDNA expression constructs. The DNA plasmids can be applied directly by mechanical inoculation or by agroinfiltration (reviewed in Annamalai & Rao, 2006). Transcription and expression of the viral RNA lead to the production of chimeric capsid proteins, which assemble into chimeric VNPs and spread systemically. Many mutant VNPs have been generated and many accumulate to titers as high as obtained for wild-type particles, as long as the mutation does not hinder capsid assembly and/or RNA packaging.

3.1.1.1 Agroinfiltration

This method exploits the plant bacterium *Agrobacterium tumefaciens*, which causes crown-gall disease in plants. *Agrobacterium tumefaciens* can invade wounded plant cells and transform the cells resulting in tumor growth. The bacteria contain a tumor-inducing plasmid, the Ti-plasmid. During infection, a segment of the Ti-plasmid is transferred into the plant cell; this segment is referred to as transfer or T-DNA. The T-DNA is incorporated into the plant genome by recombination, resulting in transient expression of the T-DNA genes. The T-DNA is flanked by 25-bp direct repeats, termed the left and right borders, that mediate the recombination event. Transfer of the T-DNA is induced by activation of the so-called virulence (vir) genes on the Ti-plasmid. Phenolic compounds, mainly acetosyringone, that are produced and released from wounded plant cells initiate expression of the vir genes. The T-DNA encodes for growth hormones, which stimulate tumor formation, as well as for unnatural amino acids such as nopaline, mannopine, and octopine, which serve as an energy source for the bacterium. The reader is referred to the following textbooks for more information on the biology of *Agrobacterium* and its use as an expression system: *Molecular Biology of the Cell* (Alberts *et al.*, 2008) and *Agrobacterium: From Biology to Biotechnology* (Tzvi & Vitaly, 2008).

In biotechnology, agrobacteria are exploited to transfer genes of interest into plant genomes for transient expression (Annamalai & Rao, 2006; D'Aoust *et al.*, 2009; Fischer *et al.*, 1999a,b; Lico *et al.*, 2008; Sheludko, 2008). The T-DNA genes are replaced with the genes of interest, such as the viral genome. Vectors used are typically binary shuttle vectors that allow replication in *Escherichia coli* as well as in agrobacteria. This facilitates convenient and fast cloning in *E. coli* prior to transforming the plasmid into agrobacteria. For inoculation of Agrobacterium (termed agroinfiltration), a suspension of the bacteria is injected into the leaves using a syringe (Fig. 3.1). To facilitate activation and transfer of the T-DNA containing the genes of interest, acetosyringone is added to the agrobacteria suspension. Two to 3 weeks post-inoculation, leaves can be harvested and analyzed, and if stably expressed, the mutant VNPs can be extracted or passaged onto subsequent plants (Lico *et al.*, 2008).

Figure 3.1 Agroinfection method. Adapted from Sainsbury, F., and Lomonossoff, G. P. (2008) Extremely high-level and rapid transient protein production in plants without the use of viral replication, *Plant Physiol.*, **148**, 1212–1218.

3.2 HETEROLOGOUS EXPRESSION SYSTEMS

Heterologous expression systems are of critical importance for large-scale production of VLPs and mutant particles. Various expression systems have been developed for many different viruses. For many years, the study of virus structure and assembly was the driving force for developing heterologous expression systems. Only a heterologous expression system allows the creation of assembly intermediates, which are essential for understanding viral assembly. Mutations that perturb virus assembly can have dramatic effects on virus viability; *in vivo* studies of assembly intermediates in the natural host are technically challenging. Heterologous expression systems offer enormous flexibility for investigating protein–protein and protein–nucleic acid interactions un-coupled from the processes of transcription, protein expression, and replication (reviewed in Schneemann & Young, 2003).

Heterologous expression systems are of great interest as they allow the production of VLPs. VLPs are devoid of infectious nucleic acids and thus cannot replicate themselves. VLPs are considered safer from an agricultural point of view and human health perspective. Heterologous expression systems used to generate VLPs include use of bacteria, yeast, insect cells, and mammalian cells. All these systems have advantages and disadvantages in terms of yield, scale, time, costs, assembly efficiency, and biological integrity (reviewed in Schneemann & Young, 2003). In the following sections, we will discuss different systems and highlight their benefits and pitfalls (summarized in Table 3.1).

Table 3.1 Summary of major expression systems for the production of VNPs and VLPs

Expression system	Vector/inoculum	Advantages	Disadvantages	Yields and time frame	VNPs/VLPs
Plants (native host)	Infected leaf material Virus particles DNA plasmid or *in vitro* transcripts	Intact particles High yield Ease of inoculation and purification Inexpensive Economic	Recombination events can lead to deletions and reversion to wild-type progeny Particles are infectious	1–2 g/1 kg leaf material 2–3 weeks 1–2 months to generate mutants	BMV CCMV CPMV HCSRV RCNMV TYMV PVX TMV
Prokaryotic *Escherichia coli* *Pseudomonas fluorescens*	Expression plasmid	High yield Commercial support (reagents and kits) Inexpensive Large-scale fermentation	Proteins frequently insoluble Coat proteins do not always assemble into particles Expression of multiple proteins challenging	1–2 g/1 l cell culture 1–2 days	CCMV TMV MS2 Qβ HK97 P22 T7
Eukaryotic Yeast: *Saccharomyces cerevisiae* *Pichia pastoris*	Expression plasmid	Assembly of intact particles High yield Ease of methods Commercial support (reagents and kits) Inexpensive Large-scale fermentation Post-translational modifications	Expression of multiple proteins is challenging Cell lysis is difficult and can lead to reduced yield	0.5 g/1 l cell culture 1–2 days	BMV CCMV Qβ
Eukaryotic Baculovirus	Recombinant baculovirus	Assembly of intact VLPs Commercial support (reagents and kits) Multiple proteins can be expressed at the same time Moderate yields Post-translational modifications	Expensive on a large scale	0.05 g/1 l cell culture Several days	FHV CPV CPMV

References are given throughout the text.

3.2.1 *E. coli* and Other Prokaryotic Expression Systems

E. coli is one of the most widely used hosts for production of heterologous proteins in general. Its genetics, biochemistry, and molecular biology are better characterized than any other microorganism. This detailed knowledge led to the development of sophisticated cloning and expression systems. Various kits are commercially available that allow manipulation, cloning, and expression with ease and convenience. The cells have a rapid doubling time (20–30 min), the growth media are cheap, and *E. coli* can be produced in large-scale fermentation systems. These features make *E. coli* a popular choice as an expression system.

Several viral coat proteins have been expressed using *E. coli*, including the coat proteins of the bacteriophages MS2, Qβ, HK97, P22, and M13, as well as the plant viruses CCMV and TMV (Gertsman *et al.*, 2009; Grieco *et al.*, 2009; Hwang *et al.*, 1994a,b; Kang *et al.*, 2008; Peabody, 2003; Sidhu, 2001; Strable *et al.*, 2008; Zhao *et al.*, 1995).

A limitation of the *E. coli* system is that eukaryotic proteins often do not assemble correctly, are insoluble, and accumulate in so-called inclusion bodies. CCMV coat proteins may be expressed to high levels in *E. coli*; however, self-assembly into particles could not be achieved (Zhao *et al.*, 1995). The produced protein must be further purified, denatured, refolded, and then assembled into the capsids *in vitro* (Section 3.5), but this is cumbersome, inefficient, and not suitable for large-scale production. To overcome these limitations, alternative expression systems have been applied. Intact CCMV VLPs were produced and self-assembled using the prokaryote *Pseudomonas fluorescens* (Phelps *et al.*, 2007b). The *P. fluorescens*-based expression system is an attractive alternative to *E. coli*; it is easy to manipulate, provides high yields of soluble and active proteins, and is suitable for large-scale fermentation.

In general, for eukaryotic viruses, it is more common to produce self-assembled VLPs using a eukaryotic expression system such as yeast or baculovirus-based expression (Sections 3.2.2 and 3.2.3).

3.2.1.1 Unnatural amino acid incorporation into VLPs using a heterologous expression system

In addition to the creation of mutant particles, expression of VLPs in heterologous expression system allows the incorporation of unnatural amino acid side chains into the viral coat proteins. This has been shown for *Hepatitis B virus* (HBV) core particles[1] and Qβ particles. Here, azide- or alkyne-containing unnatural amino acids were successfully incorporated into the viral coat proteins (Strable *et al.*, 2008). The azide or alkyne functional group allows chemical modification of the particles using Cu-catalyzed azide–alkyne cycloadditions (click chemistry), a highly specific and effective functionalization method (discussed in Chapter 4).

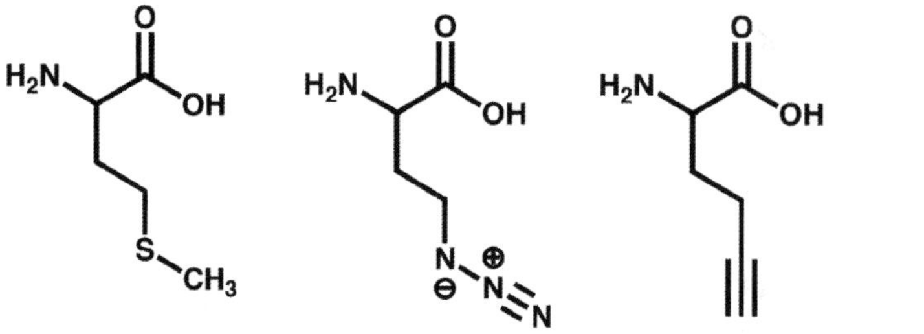

| Methionine | Azidohomoalanine | Homopropargylglycine |

Figure 3.2 Methionine and analogs azidohomoalanine and homopropargylglycine.

In brief, genetic engineering was used to insert Met codons at positions where the unnatural amino acid is desired. The modified plasmids are transformed into *E. coli.* After initial growth with natural amino acids, Met is removed from the cell media and replaced with the azide- or alkyne-containing unnatural amino acids azidohomoalanine and homopropargylglycine (Fig. 3.2). At this stage, expression of the viral coat proteins is induced, which leads to incorporation of the amino acid analogous into the capsid proteins. Using this strategy, intact virions containing novel azide- or alkyne-functionality on the capsid surface were self-assembled and isolated from the bacteria. Recombinant particles were chemically functionalized using click reactions (Strable *et al.*, 2008).

3.2.2 Yeast Expression Systems

Yeast expression systems have been widely used, and expression systems have been developed for a range of species including *Saccharomyces cerevisiae* and *Pichia pastoris*. Similar to *E. coli*, the genetics, biochemistry, and molecular biology of yeast have been extensively studied. With this knowledge in hand, various commercially available kits have been developed

---

[1] *Hepatitis B virus* infects the liver, causing inflammation that is referred to as hepatitis. HBV core particles are non-infectious particles formed during HBV infection. HBV cores or capsids are immunogenic as they consist of the surface antigen protein. HBV core particles play an important role in vaccine development.

and are available for straightforward modification, cloning, and production. Yeast production is fast, cheap, and can be scaled up using fermentation technology. The advantages of yeast over *E. coli* are that yeast cells are able to secrete soluble proteins and perform post-translational modifications, such as proteolytic processing, phosphorylation, and glycosylation.

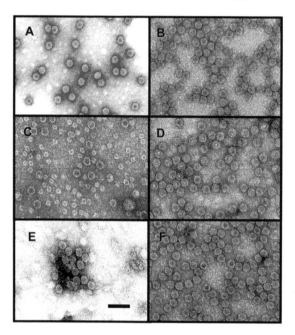

Figure 3.3 Transmission electron micrographs comparing *Cowpea chlorotic mottle virus* (CCMV) purified from plants with CCMV VLPs purified from the *P. pastoris* expression system. (A) Wild-type CCMV containing RNA purified from cowpea plants and (B) wild-type CCMV VLPs produced in the heterologous *P. pastoris* system. Mutants produced in the *P. pastoris* expression system included NΔ34 = altered interior surface charge (a range of particle sizes are observed) (C), SubE = deleted N-terminus (D), CPPep11 = peptide insertion in surface-exposed loop (E), and 81/148 = alteration of subunit interfaces at the metal binding sites (F). The scale bar is 100 nm. Reproduced with permission from Brumfield, S., Willits, D., Tang , L., Johnson , J. E., Douglas, T., and, Young, M. (2004) Heterologous expression of the modified coat protein of *Cowpea chlorotic mottle bromovirus* results in the assembly of protein cages with altered architectures and function, *J. Gen. Virol.*, **85**(Pt 4), 1049–1053.

VLPs of the following viruses have been successfully produced in yeast: the plant viruses BMV (*S. cerevisiae*) and CCMV (*P. pastoris*), as well as the bacteriophage Qβ (*S. cerevisiae* and *P. pastoris*) (Brumfield *et al.*, 2004; Freivalds *et al.*, 2006; Krol *et al.*, 1999). Intact VLPs are typically self-assembled within yeast and can be isolated in high yields by cell lysis and differential ultracentrifugation. Yields of up to 0.5 g VLP per liter cell culture can be obtained. The *Human papilloma virus* (HPV) vaccine Gardasil (VLPs of HPV) is produced in yeast expression system (Merck & Co Inc., Whitehouse Station, NJ, USA).

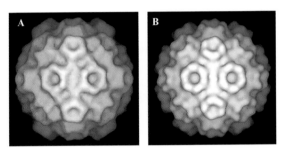

Figure 3.4 Cryo-electron microscopy reconstructions of the SubE CCMV mutant VLP produced in yeast (deleted N-terminus) (A) and wild-type CCMV produced in plants (B) showing the structural similarity between the particles. Reproduced with permission from Brumfield, S., Willits, D., Tang, L., Johnson, J. E., Douglas, T., and Young, M. (2004) Heterologous expression of the modified coat protein of *Cowpea chlorotic mottle bromovirus* results in the assembly of protein cages with altered architectures and function. *J, Gen. Virol.*, **85**(Pt 4), 1049–1053.

Yeast expression of VLPs is highly flexible and allows production of mutant particles that are unlikely to self-assemble in the natural host *in vivo*. As a proof of concept, three general classes of coat protein modifications were introduced into CCMV that altered the interior of the capsid, the interface between adjacent coat protein subunits, or the exterior capsid surface. VLPs were expressed in *P. pastoris* (Brumfield *et al.*, 2004). The modifications were designed to: (i) alter the interior surface charge of the N-terminus of the coat protein, which is the nucleic acid binding region, (ii) delete the N-terminus, (iii) insert a peptide into a surface loop, and (iv) modify interactions between viral coat protein subunits by altering the metal-binding sites. As all modifications have effects on viral assembly and nucleic acid packaging, it is unlikely that any of these mutants would assemble into intact particles in the natural host. Expression in yeast yielded VLPs that were indistinguishable from plant-derived wild-type particles. Empty VLPs as well as nucleic acid–containing particles were produced and isolated in high yields (the nucleic acid is derived from the host). Capsid proteins with altered N-termini assembled into empty particles only; and mutant with deleted N-terminus formed particles with a range of sizes (Figs 3.3 and 3.4) (Brumfield *et al.*, 2004).

3.2.3 Baculovirus-Based Expression System

Baculoviruses are insect viruses with large double-stranded DNA genomes that can accommodate multiple foreign genes of interest. Baculoviruses can be grown in cultured insect cells, such as *Spodoptera frugiperda* (Sf) lines 9 and 21 and *Trichoplusia ni* cells. The gene(s) of interest are typically placed under control of the viral polyhedrin or p10 promoters; both give rise to high yields of gene product. A limitation of the system is the large size of the baculovirus genome; unique restriction sites are not available for simple cloning. Alternative strategies employ the insertion of the gene of interest in a shuttle vector and its introduction into the expression vector by recombination. There are many commercial reagents available to facilitate baculovirus expression; however, compared with *E. coli*- or yeast-based expression systems these methods are relatively time-consuming and cumbersome. It typically takes a minimum of several days to a few weeks until the recombinant virus is constructed and a molecular clone is isolated.

The advantages of the system are reliability with regard to accumulation of soluble proteins, high probability of self-assembly of VLPs, and post-translational modifications, such as proteolytic processing, phosphorylation, and glycosylation. The glycosylation pattern can differ in insect cells compared with that observed in mammalian cells (reviewed in Schneemann & Young, 2003). For the production of eukaryotic proteins, the baculovirus-based expression has long been the system of choice. VLPs are made commercially in this system. The vaccine *Cervarix*, an HPV vaccine based on HPV VLPs, is produced in the baculovirus system and has been commercialized in 2007. Cervarix is manufactured by GlaxoSmithKline (www.gsk.com).

A   **B**

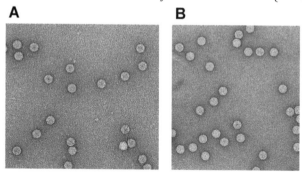

Figure 3.5 Electron micrographs of gradient-purified authentic *Flock House virus* (FHV) particles and synthetic FHV virus-like particles (VLPs). Authentic particles were isolated from FHV-infected *Drosophila melanogaster* cells (A); synthetic VLPs were isolated from recombinant baculovirus-infected *S. frugiperda* cells (B). TEM micrographs were provided by courtesy of Dr. Arno Venter and Prof. Anette Schneemann (The Scripps Research Institute, La Jolla, CA).

For nanotechnology applications, the baculovirus expression system has been successfully used to produce VLPs of the insect virus FHV, the mammalian virus CPV, and the plant virus CPMV (Saunders *et al.*, 2009;

Schneemann *et al.*, 1993; Shanks & Lomonossoff, 2000; Singh *et al.*, 2006). Intact VLPs indistinguishable from native particles were self-assembled and could be isolated in all three cases; FHV VLPs are shown in Fig. 3.5. Compared with other expression systems, yields are relatively low and the system is expensive to run on a large scale; approximately 0.05 mg VLP per 1 liter cell culture can be routinely produced.

3.2.4 Plant-Based Expression as a Heterologous System

Production of proteins, especially pharmaceutically relevant proteins, in plants is an important sector of biotechnology. Proteins can be expressed in transgenic plants, in which the protein of interest is stably incorporated into the plant genome. However, this is a time-consuming process and it can take up to a year to generate transgenic plants that are stably expressing the gene of interest. An alternative, and much faster and more flexible system is to use plant viruses as expression vectors for high-level transient expression (Awram *et al.*, 2002; Canizares *et al.*, 2005; Johnson *et al.*, 1997; Porta & Lomonossoff, 1998; Scholthof *et al.*, 1996).

The advantages of heterologous protein expression in plants are high yields, ease of purification, lack of mammalian contaminants (which is important when producing pharmaceutically relevant proteins), ease of manipulation, eukaryotic modification machinery, ease of scale-up, and cost-effectiveness. The limitations of viral vectors are the high mutation rates, which can lead to reversion to wild-type genomes and loss of the foreign gene of interest. Also there are limitations to the size of the foreign gene insert that can be tolerated by the virus.

There are several examples where VLPs have been produced using plant-based expression systems. BMV VLPs have been produced using a TMV-based expression vector (Choi & Rao, 2000). Also, CPMV- and PVX-based expression vectors have been exploited to produce HBV core particles in plants (Mechtcheriakova *et al.*, 2006).

3.4 PRODUCTION OF RECOMBINANT Ads and AAV

The previous sections dealt with the production of bacteriophages, insect viruses, and plant viruses. A few mammalian viruses have been exploited for nanotechnology, such as recombinant Ad and AAV vectors, as well as CPV, *Rubella virus,* and *Influenza* virus. To date, the only systems extensively used for biomedical nanotechnology are recombinant Ads and AAV. We will outline routes of production of recombinant Ads and AAV in this section; the reader is referred to the specialized literature regarding the production of other systems (Fischlechner *et al.*, 2005; Singh *et al.*, 2006; Toellner *et al.*, 2006).

Recombinant Ad and AAV vectors are widely used for gene therapy approaches (see Chapter 8) and to date are the most common VNPs that are under investigation in clinical trials. Large-scale production is thus an important factor. For the most part, Ad and AAV are expressed using tissue culture methods. For safer use in gene-delivery applications, replication-deficient vectors are preferable. First-generation Ad vectors (genes E1 and E3 deleted) and second-generation Ad vectors (E1, E3, and/or E4 deleted) are based on type 5 Ad and rendered replication deficient through deletion of viral proteins involved in replication and early gene regulation. Propagation of replication-deficient Ads is facilitated by co-transfection of cells with so-called helper vectors or plasmids that supply the necessary replications proteins *in trans* (in *trans* means "acting from another molecule").

AAV is a defective virus and relies by nature on the presence of helper viruses, such as Ad (serotype 5) or HSV (serotype 1) (reviewed in Lai *et al.*, 2002).

Insertion of the gene of interest can be achieved either by ligating the sequence into backbone vector fragments or through homologous recombination events. Human embryonic kidney (HEK) 293 cells are typically used for propagation of recombinant Ads or AAVs. A range of optimized propagation techniques have been developed (see the following reviews and references therein: Choi *et al.*, 2007a,b; Clark, 2002; Grieger & Samulski, 2005; Jozkowicz & Dulak, 2005; Lai *et al.*, 2002; Lu, 2004; Segura *et al.*, 2008). Also, there is substantial commercial support and a wide variety of kits and reagents are available.

Ads and AAVs grow to high titers in tissue culture and yields obtained are typically a few milligrams per 1 l cell culture. Compared with other systems, yields are only moderate. Heterologous expression of plant or bacteriophage VLPs gives rise to yields that are 10–100 times higher than those obtained for Ads and AAV. Bioreactor techniques have been developed for large-scale production (Iyer *et al.*, 1999). Yields in standard tissue culture are usually higher compared with large-scale bioreactor production; on the downside tissue culture techniques are labor-intensive and have limited potential for large-scale production.

3.5 *IN VITRO* ASSEMBLY, DISASSEMBLY, AND RE-ASSEMBLY METHODS

As mentioned above, coat proteins expressed in *E. coli* typically do not self-assemble into intact capsids; this is, for example, the case for CCMV (Zhao *et al.*, 1995). However, the proteins can be purified from *E. coli* lysates and then assembled into intact VLPs using *in vitro* assembly conditions. *In vitro* assembly methods have been established for a broad range of viruses. Intact virions can be disassembled and re-assembled *in vitro*. This is typically achieved by first exposing the particles to a disassembly buffer, whereby coat proteins can then be-reassembled by dialysis against an assembly buffer. These techniques are extensively discussed in Chapter 5 and the reader is referred to references therein.

Assembly and disassembly of viruses is well understood for some viruses, and *in vitro* assembly methods play an important role in nanotechnology. These disassembly techniques allow release of the natural cargo (nucleic acid) and are used to: (i) generate empty particles (Section 3.6) and (ii) hybrid VNPs. Hybrid VNPs or hybrid materials in general are a combination of two or more different components. For example, a VNP with an encapsulated gold core is considered a hybrid VNP consisting of the proteins shell and gold core. Re-assembly allows packaging of artificial cargos and leads to the generation of hybrid VNPs. The artificial cargo can include drug molecules, imaging moieties, synthetic polymers, or inorganic nanoparticles (discussed in detail in Chapter 5).

Re-assembly techniques also allow generation of particles with altered symmetry, for example, icosahedral particles with $T = 1$, *pseudo T = 2*, and $T = 3$ symmetries can be generated using BMV (Sun *et al.*, 2007). Not only can differently sized particles be generated, the coat proteins can also be assembled into structures, such as tubes and sheets. It is truly fascinating that a single protein, which *in planta* self-assembles into discrete, uniform, and monodisperse particles, can also be assembled into a range of different structures. This level of control and flexibility cannot be achieved with synthetic nanobuilding blocks. VNPs are intriguing materials, and understanding the self-assembly processes is expected to have significant impact on a range of technologies, including "smart" protein and/or inorganic nanoparticle design.

The self-assembly of viral rods, especially TMV, has also been extensively studied and has led to a range of applications in nanotechnology, the examples of which are discussed below.

3.5.1 BMV and CCMV

The coat proteins of the plant viruses BMV and CCMV are extremely flexible. BMV and CCMV coat proteins can assemble into *pseudo T = 2*, and $T = 3$ particles, as well as into sheets, stacks, and tubular structures (Bancroft *et al.*, 1967; Krol *et al.*, 1999; Pfeiffer *et al.*, 1976; Pfeiffer & Hirth, 1974). BMV particles have also been observed in a $T = 1$ geometry; $T = 1$ particles are formed *in vitro* after cleavage of the N-terminal 63 or 35 amino acid residues of the coat protein by an endogenous protease or by trypsin. $T = 3$ particles could also be directly condensed into $T = 1$ particles without undergoing complete disassembly (Fig. 3.6) (Larson *et al.*, 2005; Lucas *et al.*, 2001).

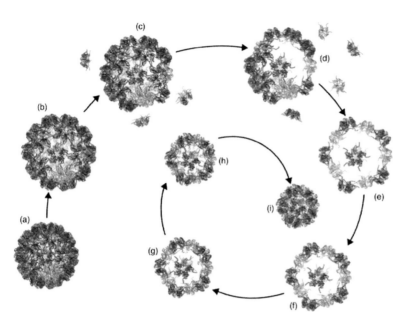

Figure 3.6 A diagram showing the sequence of events in the transformation of a native $T = 3$ *Brome mosaic virus* (BMV) virion in panel A into a $T = 1$ particle (I). As a consequence of elevated pH and high salt concentration, the native virion in panel A undergoes a transition to the swollen form in panel B, thereby allowing exposure of the virion interior to proteases and ribonucleases. As proteolytic and nucleolytic cleavage proceeds, RNA fragments, amino terminal polypeptides, and hexameric units from the capsid are lost (C–E). Since hexamers are lost, restructuring begins (E) when pentameric units start their 378 rotation, as they condense to the smaller particle. In panels F–H, the pentamers continue their rotation as contraction of the assembly proceeds simultaneously. Reproduced with permission from Larson ; S. B., Lucas , R. W., and McPherson , A. (2005) Crystallographic structure of the $T = 1$ particle of brome mosaic virus, *J. Mol. Biol.*, **346**(3), 815–831.

3.5.2 TMV

The assembly and disassembly of TMV coat proteins has been extensively studied; reconstitution experiments yield either disk- or rod-shaped structures as shown in Fig. 3.7 (Fraenkel-Conrat & Williams, 1955). When disassembled and exposed to physiological conditions the coat proteins of TMV form disk structures. The proteins aggregate into a two-layer cylindrical structure, each layer consisting of 17 coat protein monomers (16 1/3 molecules are present in each turn of the assembled helix). By lowering the pH and exposure to its nucleic acid, assembly into intact virions can be initiated. Assembly of TMV initiates at the RNA origin of assembly (OAS) site, which is located at 75–432 nt on the viral RNA. Elongation of the coat proteins occurs in 5′→3′ and 3′→5′ direction; assembly toward the 3′ direction is considerably slower (Butler & Klug, 1972; Culver, 2002; Fraenkel-Conrat & Williams, 1955; Klug, 1972, 1999).

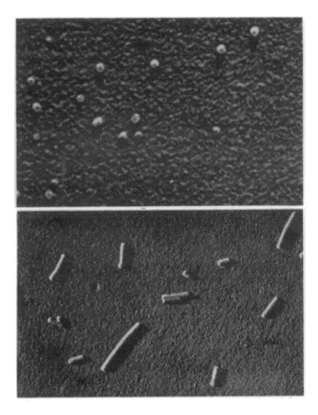

Figure 3.7 Electron micrographs of *Tobacco mosaic virus* (TMV) proteins used in reconstitution experiments. Top panel shows disk-shaped particles with central holes. Bottom panel shows rod-shaped particles with morphology identical to that of normal TMV. Reproduced from Fraenkel-Conrat , H., and Williams , R. C. (1955) Reconstitution of active tobacco mosaic virus from its inactive protein and nucleic acid components, *Proc. Natl. Acad. Sci. USA*, **41**(10), 690–698.

Coat protein monomers can also be self-assembled into rods without nucleic acid. These methods have been exploited to fabricate self-assembled light harvesting systems. Different fluorescent dye-labeled coat proteins were self-assembled into either disks or rods. During this process, the different fluorescent dyes (i.e., chromophores) are brought into near proximity, allowing energy transfer between the chromophores (Miller *et al.*, 2007). This system may be further developed for the construction of a photovoltaic device and is discussed in Chapter 4.

Different materials applications have made use of the polarity of TMV rods and the fact that the particles can be partially disassembled at one end or the other. The helical encapsulation of the RNA molecule results in sequence-definable 5′ and 3′ ends. Partial disassembly to expose the RNA molecule at either end can be achieved. These methods were extensively studied in the 1970s, and the reaction conditions to achieve partial disassembly from 5′→3′ and 3′→5′, respectively, of the viral rod are known (Ohno & Okada, 1977; Wilson, 1978). To achieve stripping from the 5′ end mild sodium dodecyl sulfate (SDS) treatment can be used. Disassembly from the 3′ end is typically achieved using a mild dimethyl sulfoxide (DMSO) treatment.

These methods have been exploited to specifically immobilize TMV on solid supports. Spatially oriented assembly of partially disassembled TMV with exposed RNA was achieved in a controlled manner. TMV was disassembled selectively on the 5′ end. The exposed RNA molecules were then utilized for immobilization of the particles via nucleic acid hybridization using complementary oligonucleotides bound on solid supports (Yi *et al.*, 2005). Furthermore, by using differentially labeled TMV particles and a micropatterned substrate, the construction of a patterned TMV microarray was achieved (Yi *et al.*, 2007). Immobilization and multilayer array fabrication techniques are discussed in detail in Chapter 7.

3.6 GENERATION OF EMPTY AND NON-INFECTIOUS PARTICLES

With developments in viral nanotechnology, there has been increasing demand for empty particles or non-infectious particles. VLPs expressed in heterologous expression systems are non-infectious as they lack genomic nucleic acids; however, they typically pack random host nucleic acids.

Various *in vitro* techniques have been developed that allow release and removal of nucleic acids from capsids to generate particles that are truly empty. These methods have also been adapted to packaging artificial cargos into capsids (discussed in Chapter 5).

Empty particles, for example, can be self-assembled *in vitro* from TMV or CCMV coat proteins (Gillitzer *et al.*, 2006; Miller *et al.*, 2007). Alternatively, nucleic acid can be released by making use of the pH-dependent swelling mechanism of particles, such as CCMV and HCRSV (Douglas & Young, 1998; Ren *et al.*, 2006). At pH 8.0, the particles undergo structural changes and appear in a swollen and open conformation. The encapsidated nucleic acid, which is RNA in both cases, can then be degraded by alkaline hydrolysis and removed by differential ultracentrifugation. Mild denaturants, such as urea or RNase (enzymatic degradation of the RNA) can also be added to ensure degradation of the nucleic acids.

Alkaline hydrolysis is also an effective method for particles that do not undergo extensive swelling, such as CPMV. Alkaline hydrolysis can be applied either to purified CPMV particles (Ochoa *et al.*, 2006) or during the initial steps of the extraction protocol (Phelps *et al.*, 2007a). The reported shelf-life of empty particles generated by alkaline hydrolysis is about 2 weeks (Ochoa *et al.*, 2006), which limits potential long-term applications.

The nucleic acids of T7 phage can also be extracted by alkaline hydrolysis (Liu *et al.*, 2006); nucleic acid extraction via osmotic shock has been applied as an alternative method (Liu *et al.*, 2005).

Some viruses form empty particles during the natural infection process. CPMV particles, for example, form empty as well as nucleic acid-containing particles *in planta*. CPMV particles form three different components that have identical protein composition but differ in their RNA contents (Bancroft, 1962; Bruening & Agrawal, 1967; Wu & Bruening, 1971); these components can be separated by density (Fig. 2.10). The particles representing the top component are devoid of RNA, whereas the middle and bottom components each contain a single RNA molecule, RNA-2 and RNA-1, respectively (Lomonossoff & Johnson, 1991). However, the yield of empty particles in a wild-type virus preparation is rather small (<10 %). Empty, so-called ghost T7 phages can also be isolated (in low yields) at the early stage of infection (Studier, 1972).

To address the need for non-infectious particles for safe use in materials and medical applications, scientists have also developed methods to render intact, genome-containing particles non-infectious. Short-wave (254 nm) UV irradiation using a dosage of 2.0–2.5 J/cm^2 has been demonstrated as an efficient strategy to crosslink the RNA genome within intact CPMV particles. The particles were non-infectious for plants, remained intact, and maintained chemical reactivity and cellular binding properties (Rae *et al.*, 2008). UV inactivation of CPMV prior to or after modification thus provides a route for designing non-infectious particles for safe use in *in vivo* applications or as materials. UV inactivation is a common technique and has been applied to a wide range of viral pathogens (reviewed in Lytle & Sagripanti, 2005).

3.7 "SMART" DESIGN OF MUTANT VNPs

An advantage of VNPs and other proteinaceous nanomaterials over any synthetic material is that the biological material allows tuning using either genetics or chemistry. Chemical modification strategies can also be applied to synthetic materials, although high level of spatial control can typically only be achieved using biomolecules. The genomes of viruses are small, and the nucleotide sequences of many viruses are known; in addition, infectious clones have been generated that allow genetic manipulation, cloning, and expression of a wide range of mutant particles.

A large variety of mutant VNPs have been generated for various applications; these mutants will be mentioned and discussed in detail throughout the chapters of this book. In this section, some of the basic principles of "smart" mutant design are highlighted. Many virus capsids have been crystallized and the structure has been solved to near atomic resolution; with this knowledge in hand, mutants can be designed with atomic precision. The availability of the structural data allows analyzing the surface coordinates of the particles for potentially reactive amino acid side chains. If no such sites are available, amino acids can be introduced by means of genetics. The site of insertion can be identified using the structure. It also allows modeling the presentation and orientation of peptides, proteins, or other ligands that were attached to amino acid side chains by means of chemistry.

3.8 SEVERAL GENERAL CLASSES OF MUTATIONS

1. **Mutations that add novel amino acid side chains to the capsid surface in order to allow for bioconjugation protocols.** Common examples are Cys-added mutants. Cys side chains are rarely found on the solvent-exposed exterior capsid surface; however, the thiol group of Cys residues is a useful tool for bioconjugation, as the thiol allows facile coupling to a range of metals or maleimide-activated compounds. Cys-added mutants have been generated, for example, for MS2, CCMV, CPMV, and TMV (Klem *et al.*, 2003; Miller *et al.*, 2007; Peabody, 2003; Wang *et al.*, 2002). Lys side chains are also highly versatile groups for chemical modification and have been introduced into the capsids of FHV and TMV, both of which do not naturally display any reactive Lys side chains in their coat protein sequence (Demir & Stockwell, 2002; Portney *et al.*, 2005).

2. **Mutations that add novel amino acid side chains as affinity tags to allow for immobilization or binding with and of other materials.** Hexa-Histidine tags (that is six constitutive His side chains) are commonly used as affinity tags for purification of proteins using Ni-NTA affinity columns. His tags have been introduced into VNPs, and particle purification over Ni-NTA columns could be carried out (Chatterji *et al.*, 2005). His tags have also been used as anchor groups to allow for labeling; for example, decoration of His-tagged P22 and CPMV with Ni-conjugated nanogold was shown (Chatterji *et al.*, 2005; Kang *et al.*, 2008). The Hexa-

His tag can also be used to facilitate immobilization and assembly of array structures as demonstrated for CPMV and Qβ (Medintz *et al.*, 2005; Udit *et al.*, 2008).

3. **Deletions of amino acid side chains to facilitate higher level of control over the number and sites available for bioconjugation.** If a particle offers more attachment site than desired, sites can be removed by mutagenesis. This has been demonstrated for CPMV, which has 300 addressable Lys side chains, five on each asymmetric unit. The asymmetric unit of CPMV consists of one copy each of the S and L proteins (see Fig. 2.6). Systematic deletion of Lys side chains resulted in Lys-minus mutant particles with uniquely reactive sites remaining (Chatterji *et al.*, 2004).

4. **Alteration of surface charge to allow interactions with materials via electrostatic interactions.** An example is the alteration of the surface charge of the N-terminus of the CCMV coat protein. The N-terminus is involved in nucleic acid binding and packing and is highly positively charged. Amino acid side chains were exchanged to generate an overall negatively charged N-terminus. The electrostatically altered VNP catalyzed the oxidation of Fe(II), leading to the formation of spatially constrained iron oxide nanocrystals within the interior cavity of the CCMV particles (Douglas *et al.*, 2002).

5. **Insertions of short peptide sequences or whole proteins.** A whole range of short peptide sequences and intact proteins have been introduced into viral capsids for various applications. A number of examples are listed here; the reader is referred to the respective chapters to learn about the various chimeras. Antigenic peptides have been introduced for vaccine development (Chapter 8). Targeting peptides or proteins are used to deliver VNPs to specific molecular receptors (Chapter 8), and various peptides that promote the nucleation of inorganic materials have been incorporated into VNPs to facilitate their mineralization and metallization for applications in materials science (Chapter 6).

References

Ahlquist, P., French, R., Janda, M., and Loesch-Fries, L. S. (1984) Multicomponent RNA plant virus infection derived from cloned viral cDNA, *Proc. Natl. Acad. Sci. USA*, **81**(22), 7066–7070.

Ahlquist, P., and Janda, M. (1984) cDNA cloning and *in vitro* transcription of the complete brome mosaic virus genome, *Mol. Cell. Biol.*, **4**(12), 2876–2882.

Alberts, B., Johnson, A., Lewis, J., Raff, M., Roberts, K., and Walter, P. (2008) *Molecular Biology of the Cell*, 5th edn, Garland Science, Taylor & Francis Group, New York.

Allison, R. F., Janda, M., and Ahlquist, P. (1988) Infectious *in vitro* transcripts from cowpea chlorotic mottle virus cDNA clones and exchange of individual RNA components with brome mosaic virus, *J. Virol.*, **62**(10), 3581–3588.

Annamalai, P., and Rao, A. L. (2006) Delivery and expression of functional viral RNA genomes in planta by agroinfiltration, *Curr. Protoc. Microbiol.*, Chapter 16: Unit16B 12.

Awram, P., Gardner, R. C., Forster, R. L., and Bellamy, A. R. (2002) The potential of plant viral vectors and transgenic plants for subunit vaccine production, *Adv. Virus Res.*, **58**, 81–124.

Bancroft, J. B. (1962) Purification and properties of bean pod mottle virus and associated centrifugal and electrophoretic components, *Virology*, **16**, 419–427.

Bancroft, J. B., Hills, G. J., and Markham, R. (1967) A study of the self-assembly process in a small spherical virus. Formation of organized structures from protein subunits in vitro, . *Virology*, **31**(2), 354–379.

Bruening, G., and Agrawal, H. O. (1967) Infectivity of a mixture of cowpea mosaic virus ribonucleoprotein components, *Virology*, **32**(2), 306–320.

Brumfield, S., Willits, D., Tang, L., Johnson, J. E., Douglas, T., and Young, M. (2004) Heterologous expression of the modified coat protein of *Cowpea chlorotic mottle bromovirus* results in the assembly of protein cages with altered architectures and function, *J. Gen. Virol.*, **85**(Pt 4), 1049–1053.

Butler, P. J., and Klug, A. (1972) Assembly of tobacco mosaic virus *in vitro*: effect of state of polymerization of the protein component, *Proc. Natl. Acad. Sci. USA,* **69**(10), 2950–2953.

Canizares, M. C., Nicholson, L., and Lomonossoff, G. P. (2005) Use of viral vectors for vaccine production in plants, *Immunol. Cell. Biol.*, **83**(3), 263–270.

Chatterji, A., Ochoa, W., Paine, M., Ratna, B. R., Johnson, J. E., and Lin, T. (2004) New addresses on an addressable virus nanoblock: uniquely reactive Lys residues on cowpea mosaic virus, *Chem. Biol.*, **11**(6), 855–863.

Chatterji, A., Ochoa, W. F., Ueno, T., Lin, T., and Johnson, J. E. (2005) A virus-based nanoblock with tunable electrostatic properties, *Nano Lett.*, **5**(4), 597–602.

Choi, V. W., Asokan, A., Haberman, R. A., and Samulski, R. J. (2007a) Production of recombinant adeno-associated viral vectors, *Curr. Protoc. Hum. Genet.*, Chapter 12: Unit 12 19.

Choi, V. W., Asokan, A., Haberman, R. A., and Samulski, R. J. (2007b) Production of recombinant adeno-associated viral vectors for *in vitro* and in vivo use, *Curr. Protoc. Mol. Biol.*, Chapter 16: Unit 16 25.

Choi, Y. G., and Rao, A. L. (2000) Packaging of tobacco mosaic virus subgenomic RNAs by Brome mosaic virus coat protein exhibits RNA controlled polymorphism, *Virology*, **275**(2), 249–257.

Clark, K. R. (2002) Recent advances in recombinant adeno-associated virus vector production, *Kidney Int.*, **61**(1 Suppl.), S9–S15.

Culver, J. N. (2002) Tobacco mosaic virus assembly and disassembly: determinants in pathogenicity and resistance, *Annu. Rev. Phytopathol.*, **40**, 287–308.

D'Aoust, M. A., Lavoie, P. O., Belles-Isles, J., Bechtold, N., Martel, M., and Vezina, L. P. (2009) Transient expression of antibodies in plants using syringe agroinfiltration, *Methods Mol. Biol.*, **483**, 41–50.

Demir, M., and Stockwell, M. H. B. (2002) A chemoselective biomolecular template for assembling diverse nanotubular materials, *Nanotechnology*, **13**, 541–544.

Dessens, J. T., and Lomonossoff, G. P. (1993) Cauliflower mosaic virus 35S promoter-controlled DNA copies of cowpea mosaic virus RNAs are infectious on plants, *J. Gen. Virol.*, **74**(Pt 5), 889–892.

Douglas, T., Strable, E., and Willits, D. (2002) Protein engineering of a viral cage for constrained material synthesis, *Adv. Mater.*, **14**, 415–418.

Douglas, T., and Young, M. (1998) Host–guest encapsulation of materials by assembled virus protein cages, *Nature*, **393**, 152–155.

Fischer, R., Drossard, J., Commandeur, U., Schillberg, S., and Emans, N. (1999a) Towards molecular farming in the future: moving from diagnostic protein and antibody production in microbes to plants, *Biotechnol. Appl. Biochem.*, **30** (Pt 2), 101–108.

Fischer, R., Liao, Y. C., Hoffmann, K., Schillberg, S., and Emans, N. (1999b) Molecular farming of recombinant antibodies in plants, *Biol. Chem.*, **380**(7–8), 825–839.

Fischlechner, M., Zschornig, O., Hofmann, J., and Donath, E. (2005) Engineering virus functionalities on colloidal polyelectrolyte lipid composites, *Angew Chem. Int. Ed. Engl*, **44**(19), 2892–2895.

Fraenkel-Conrat, H., and Williams, R. C. (1955) Reconstitution of active tobacco mosaic virus from its inactive protein and nucleic acid components, *Proc. Natl. Acad. Sci. USA*, **41**(10), 690–698.

Freivalds, J., Dislers, A., Ose, V., Skrastina, D., Cielens, I., Pumpens, P., Sasnauskas, K., and Kazaks, A. (2006) Assembly of bacteriophage Qbeta virus-like particles in yeast *Saccharomyces cerevisiae* and *Pichia pastoris, J. Biotechnol.*, **123**(3), 297–303.

Gertsman, I., Gan, L., Guttman, M., Lee, K., Speir, J. A., Duda, R. L., Hendrix, R. W., Komives, E. A., and Johnson, J. E. (2009) An unexpected twist in viral capsid maturation, *Nature*, **458**(7238), 646–650.

Gillitzer, E., Suci, P., Young, M., and Douglas, T. (2006) Controlled ligand display on a symmetrical protein-cage architecture through mixed assembly, *Small*, **2**(8–9), 962–966.

Grieco, S. H., Lee, S., Dunbar, W. S., Macgillivray, R. T., and Curtis, S. B. (2009) Maximizing filamentous phage yield during computer-controlled fermentation, *Bioprocess Biosyst. Eng.*, 32, 773–779.

Grieger, J. C., and Samulski, R. J. (2005) Adeno-associated virus as a gene therapy vector: vector development, production and clinical applications, *Adv. Biochem. Eng. Biotechnol.*, **99**, 119–145.

Hwang, D. J., Roberts, I. M., and Wilson, T. M. (1994a) Assembly of tobacco mosaic virus and TMV-like pseudovirus particles in *Escherichia coli. Arch. Virol. Suppl.*, **9**, 543–558.

Hwang, D. J., Roberts, I. M., and Wilson, T. M. (1994b) Expression of tobacco mosaic virus coat protein and assembly of pseudovirus particles in *Escherichia coli, Proc. Natl. Acad. Sci. U.S.A.*, **91**(19), 9067–9071.

Iyer, P., Ostrove, J. M., and Vacante, D. (1999) Comparison of manufacturing techniques for adenovirus production, *Cytotechnology*, **30**(1–3), 169–172.

Johnson, J., Lin, T., and Lomonossoff, G. (1997) Presentation of heterologous peptides on plant viruses: genetics, structure, and function, *Annu. Rev. Phytopathol.*, **35**, 67–86.

Jozkowicz, A., and Dulak, J. (2005) Helper-dependent adenoviral vectors in experimental gene therapy, *Acta Biochim. Pol.*, **52**(3), 589–599.

Kang, S., Lander, G. C., Johnson, J. E., and Prevelige, P. E. (2008) Development of bacteriophage p22 as a platform for molecular display: genetic and chemical modifications of the procapsid exterior surface. , *ChemBioChem*, **9**(4), 514–518.

Klem, M. T., Willits, D., Young, M., and Douglas, T. (2003) 2-D array formation of genetically engineered viral cages on Au surfaces and imaging by atomic force microscopy, *J. Am. Chem. Soc.*, **125**(36), 10806–10807.

Klug, A. (1972) Assembly of tobacco mosaic virus, *Fed. Proc.* **31**(1), 30–42.

Klug, A. (1999) The tobacco mosaic virus particle: structure and assembly, *Philos. Trans. R. Soc. Lond. B. Biol. Sci.*, **354**(1383), 531–535.

Krol, M. A., Olson, N. H., Tate, J., Johnson, J. E., Baker, T. S., and Ahlquist, P. (1999) RNA-controlled polymorphism in the in vivo assembly of 180-subunit and 120-subunit virions from a single capsid protein, *Proc. Natl. Acad. Sci. USA*, **96**(24), 13650–13655.

Lai, C. M., Lai, Y. K., and Rakoczy, P. E. (2002) Adenovirus and adeno-associated virus vectors, *DNA Cell. Biol.*, **21**(12), 895–913.

Larson, S. B., Lucas, R. W., and McPherson, A. (2005) Crystallographic structure of the T = 1 particle of brome mosaic virus, *J. Mol. Biol.*, **346**(3), 815–831.

Lico, C., Chen, Q., and Santi, L. (2008) Viral vectors for production of recombinant proteins in plants, *J. Cell. Physiol.*, **216**(2), 366–377.

Liu, C. M., Chung. S.-H., Jin, Q., Sutton, A., Yan, F., Hoffmann, A., Kay, B. K., Bader, S. D., Makowski, L., and Chen, L. (2006) Magnetic viruses via nano-capsid templates, *J. Magn. Magn. Mater.*, **302**, 47–51.

Liu, C. M., Jin, Q., Sutton, A., and Chen, L. (2005) A novel fluorescent probe: europium complex hybridized T7 phage, *Bioconjug. Chem.*, **16**(5), 1054–1057.

Lomonossoff, G. P., and Johnson, J. E. (1991) The synthesis and structure of comovirus capsids. *Prog. Biophys. Mol. Biol.*, **55**(2), 107–137.

Lu, Y. (2004) Recombinant adeno-associated virus as delivery vector for gene therapy —a review, *Stem Cells Dev.* **13**(1), 133–145.

Lucas, R. W., Kuznetsov, Y. G., Larson, S. B., and McPherson, A. (2001) Crystallization of Brome mosaic virus and T = 1 Brome mosaic virus particles following a structural transition, *Virology*, **286**(2), 290–303.

Lytle, C. D., and Sagripanti, J. L. (2005) Predicted inactivation of viruses of relevance to biodefense by solar radiation, *J. Virol.*, **79**(22), 14244–14252.

Mechtcheriakova, I. A., Eldarov, M. A., Nicholson, L., Shanks, M., Skryabin, K. G., and Lomonossoff, G. P. (2006) The use of viral vectors to produce hepatitis B virus core particles in plants, *J. Virol. Methods*, **131**(1), 10–15.

Medintz, I. L., Sapsford, K. E., Konnert, J. H., Chatterji, A., Lin, T., Johnson, J. E., and Mattoussi, H. (2005) Decoration of discretely immobilized cowpea mosaic virus with luminescent quantum dots, *Langmuir*, **21**(12), 5501–5510.

Mekuria, T., Bamunusinghe, D., Payton, M., and Verchot-Lubicz, J. (2008) Phloem unloading of potato virus X movement proteins is regulated by virus and host factors, *Mol. Plant Microbe Interact.*, **21**(8), 1106–1117.

Meshi, T., Ishikawa, M., Motoyoshi, F., Semba, K., and Okada, Y. (1986) *In vitro* transcription of infectious RNAs from full-length cDNAs of tobacco mosaic virus, *Proc. Natl. Acad. Sci. USA*, **83**(14), 5043–5047.

Miller, R. A., Presley, A. D., and Francis, M. B. (2007) Self-assembling light-harvesting systems from synthetically modified tobacco mosaic virus coat proteins, *J. Am. Chem. Soc.*, **129**(11), 3104–3109.

Ochoa, W. F., Chatterji, A., Lin, T., and Johnson, J. E. (2006) Generation and structural analysis of reactive empty particles derived from an icosahedral virus, *Chem. Biol.*, **13**(7), 771–778.

Ohno, T., and Okada, Y. (1977) Polarity of stripping of tobacco mosaic virus by alkali and sodium dodecyl sulfate, *Virology*, **76**(1), 429–432.

Peabody, D. S. (2003) A viral platform for chemical modification and multivalent display, *J. Nanobiotechnol.*, **1**(1), 5.

Pfeiffer, P., Herzog, M., and Hirth, L. (1976) RNA viruses: stabilization of brome mosaic virus, *Philos. Trans. R. Soc. Lond. B Biol. Sci.*, **276**(943), 99–107.

Pfeiffer, P., and Hirth, L. (1974) Aggregation states of brome mosaic virus protein, *Virology*, **61**(1), 160–167.

Phelps, J. P., Dang, N., and Rasochova, L. (2007a) Inactivation and purification of cowpea mosaic virus-like particles displaying peptide antigens from *Bacillus anthracis, J. Virol. Methods*, **141**(2), 146–153.

Phelps, J. P., Dao, P., Jin, H., and Rasochova, L. (2007b) Expression and self-assembly of cowpea chlorotic mottle virus-like particles in *Pseudomonas fluorescens, J. Biotechnol.*, **128**(2), 290–296.

Porta, C., and Lomonossoff, G. P. (1998) Scope for using plant viruses to present epitopes from animal pathogens, *Rev. Med. Virol.*, **8**(1), 25–41.

Portney, N. G., Singh, K., Chaudhary, S., Destito, G., Schneemann, A., Manchester, M., and Ozkan, M. (2005) Organic and inorganic nanoparticle hybrids, *Langmuir*, **21**(6), 2098–2103.

Rae, C., Koudelka, K. J., Destito, G., Estrada, M. N., Gonzales, M. J., and Manchester, M. (2008) Chemical addressability of ultraviolet-inactivated viral nanoparticles (VNPs). *PLoS One*, 3(10), e3315.

Ren, Y., Wong, S. M., and Lim, L. Y. (2006) *In vitro*-reassembled plant virus-like particles for loading of polyacids, *J. Gen. Virol.*, **87**(Pt 9), 2749–2754.

Roberts, A. G., Cruz, S. S., Roberts, I. M., Prior, D., Turgeon, R., and Oparka, K. J. (1997) Phloem unloading in sink leaves of *nicotiana benthamiana*: comparison of a fluorescent solute with a fluorescent virus, *Plant Cell*, **9**(8), 1381–1396.

Sainsbury, F., and Lomonossoff, G. P. (2008) Extremely high-level and rapid transient protein production in plants without the use of viral replication, *Plant Physiol.*, 148, 1212–1218.

Saunders, K., Sainsbury, F., and Lomonossoff, G. P. (2009) Efficient generation of cowpea mosaicvirus empty virus-like particles by the proteolytic processing of precursors in insect cells and plants, *Virology*, 393(2), 329–337.

Schneemann, A., Dasgupta, R., Johnson, J. E., and Rueckert, R. R. (1993) Use of recombinant baculoviruses in synthesis of morphologically distinct viruslike particles of flock house virus, a nodavirus, *J. Virol.*, **67**(5), 2756–2763.

Schneemann, A., and Young, M. J. (2003) Viral assembly using heterologous expression systems and cell extracts, *Adv. Protein Chem.*, **64**, 1–36.

Scholthof, H. B., Scholthof, B. G., and Jackson, A. O. (1996) Plant virus vectors for transient expression of foreign proteins in plants, *Annu. Rev. Phytopathol.*, **34**, 229–323.

Segura, M. M., Alba, R., Bosch, A., and Chillon, M. (2008) Advances in helper-dependent adenoviral vector research, *Curr. Gene Ther.*, **8**(4), 222–235.

Shanks, M., and Lomonossoff, G. P. (2000) Co-expression of the capsid proteins of Cowpea mosaic virus in insect cells leads to the formation of virus-like particles, *J. Gen. Virol.*, **81**(Pt 12), 3093–3097.

Sheludko, Y. V. (2008) Agrobacterium-mediated transient expression as an approach to production of recombinant proteins in plants, *Recent Pat. Biotechnol.*, **2**(3), 198–208.

Sidhu, S. S. (2001) Engineering M13 for phage display, *Biomol. Eng.*, **18**(2), 57–63.

Silva, M. S., Wellink, J., Goldbach, R. W., and van Lent, J. W. (2002) Phloem loading and unloading of *Cowpea mosaic virus* in *Vigna unguiculata, J. Gen. Virol.*, **83**(Pt 6), 1493–1504.

Singh, P., Destito, G., Schneemann, A., and Manchester, M. (2006) Canine parvovirus-like particles, a novel nanomaterial for tumor targeting, *J. Nanobiotechnol.*, **4**, 2.

Strable, E., Prasuhn, D. E., Jr., Udit, A. K., Brown, S., Link, A. J., Ngo, J. T., Lander, G., Quispe, J., Potter, C. S., Carragher, B., Tirrell, D. A., and Finn, M. G. (2008) Unnatural amino acid incorporation into virus-like particles, *Bioconjug. Chem.*, **19**(4), 866–875.

Studier, F. W. (1972) Bacteriophage T7 *Science*, **176**(33), 367–376.

Sun, J., DuFort, C., Daniel, M. C., Murali, A., Chen, C., Gopinath, K., Stein, B., De, M., Rotello, V. M., Holzenburg, A., Kao, C. C., and Dragnea, B. (2007) Core-controlled polymorphism in virus-like particles, *Proc. Natl. Acad. Sci. USA,* **104**(4), 1354–1359.

Toellner, L., Fischlechner, M., Ferko, B., Grabherr, R. M., and Donath, E. (2006) Virus-coated layer-by-layer colloids as a multiplex suspension array for the detection and quantification of virus-specific antibodies, *Clin. Chem.*, **52**(8), 1575–1583.

Tzvi, T., and Vitaly, C. (2008) *Agrobacterium: From Biology to Biotechnology*, Springer, New York.

Udit, A. K., Brown, S., Baksh, M. M., and Finn, M. G. (2008) Immobilization of bacteriophage Qbeta on metal-derivatized surfaces via polyvalent display of hexahistidine tags, *J. Inorg. Biochem.*, **102**(12), 2142–2146.

Wang, Q., Lin, T., Johnson, J. E., and Finn, M. G. (2002) Natural supramolecular building blocks: cysteine-added mutants of cowpea mosaic virus, *Chem. Biol.*, **9**(7), 813–819.

Wilson, T. M. (1978) The polarity of stripping of coat protein subunits from the RNA in tobacco mosaic virus by dimethylsulphoxide, *FEBS Lett.*, **87**(1), 17–20.

Wu, G. J., and Bruening, G. (1971) Two proteins from cowpea mosaic virus, *Virology*, **46**(3), 596–612.

Xiong, Z. G., and Lommel, S. A. (1991) Red clover necrotic mosaic virus infectious transcripts synthesized *in vitro, Virology*, **182**(1), 388–392.

Yi, H., Nisar, S., Lee, S. Y., Powers, M. A., Bentley, W. E., Payne, G. F., Ghodssi, R., Rubloff, G. W., Harris, M. T., and Culver, J. N. (2005) Patterned assembly of genetically modified viral nanotemplates via nucleic acid hybridization, *Nano Lett.*, **5**(10), 1931–1936.

Yi, H., Rubloff, G. W., and Culver, J. N. (2007) TMV microarrays: hybridization-based assembly of DNA-programmed viral nanotemplates, *Langmuir*, **23**(5), 2663–2667.

Zhao, X., Fox, J. M., Olson, N. H., Baker, T. S., and Young, M. J. (1995) *In vitro* assembly of cowpea chlorotic mottle virus from coat protein expressed in *Escherichia coli* and in vitro-transcribed viral cDNA, *Virology*, **207**(2), 486–494.

Chapter 4

THE ART OF BIOCONJUGATION: FUNCTIONALIZATION OF VNPs

Bioconjugation allows the linking of two or more molecules together to create a novel hybrid material, resulting in the introduction of a moiety (a functional group) into a biomolecule. Bioconjugation techniques have been applied to nearly every class of biomolecule, such as lipids, carbohydrates, polysaccharides, nucleic acids, oligonucleotides, peptides, proteins, antibodies, and viral nanoparticles (VNPs).

A large range of commercially available reagents have been developed; for example, see the online catalogs of Pierce (http:www.piercenet.com), Invitrogen (http://www.invitrogen.com), and Solulink (http://www.solulink. com). Commercially available reagents range from small organic fluorescent dyes, which can be used as tracers and imaging molecules, to bi- and trivalent chemical linkers that allow combining two or three molecules.

Labeling techniques find applications in nearly any discipline. In *basic research*, derivatization techniques are used for the analysis of protein structure and function, and protein purification, and a large range of labeled conjugates are used for immunohistochemistry. In *medicine*, labeled proteins and nanomaterials are used in *diagnostics*, *imaging*, and *therapy*. Furthermore, immobilized nanomaterials, enzymes, antibodies, and nucleic acids find applications as biosensors and tracers in assays used for *environmental*, *industrial*, *military*, and *clinical testing*.

Bioconjugation techniques play an important role in viral nanotechnology. For many applications, the VNP is used as a building block or construction material to generate a functional entity. For medical applications, for example, VNPs are labeled with imaging molecules such as fluorescent dyes or contrast agents such as gadolinium, allowing detection by fluorescence microscopy or magnetic resonance imaging (MRI). Also, targeting ligands, such as peptides, proteins, or antibodies, as well as

Viral Nanoparticles: *Tools for Materials Science and Biomedicine*
By Nicole F. Steinmetz and Marianne Manchester
Copyright © 2011 by Pan Stanford Publishing Pte. Ltd.
www.panstanford.com

therapeutic molecules can be linked to VNPs in order to generate "smart" devices for targeted drug delivery (discussed in Chapter 8). For potential applications in novel materials, conducting materials have been chemically linked to VNPs to create nanocircuits, which may be useful for novel nanoelectronic devices. VNPs have also been applied as sensors and have been developed for light-harvesting systems (discussed in Section 4.4). Nearly any chemical compound or functional molecule can be covalently linked to VNPs; hence, the potential applications of VNPs are open-ended.

This chapter will provide an overview of the different bioconjugation chemistries that have been applied to a variety of VNP building blocks. These protocols range from standard labeling protocols using commercially available compounds to advanced chemistries such as click chemistry and diazonium coupling strategies. For general background reading on bioconjugation techniques, the following textbooks are recommended: *Bioconjugate Techniques* (Hermanson, 1996) and *Bioconjugation* (Aslam & Dent, 1999).

4.1 OVERVIEW: ADDRESSABLE SURFACE GROUPS

The capsids of VNPs are composed of many identical copies of coat proteins, which consist of amino acids. Amino acids serve as targets for bioconjugation chemistry. Some viruses are glycosylated and the carbohydrates may also be utilized for functionalization (see Section 4.1.5).

There are 20 common amino acids found in proteins (Fig. 4.1). Each amino acid contains an amino group and a carboxylate attached to a central carbon atom, the α-carbon, and a side chain. The side chain is different and characteristic for each of the 20 common amino acids. The amino and carboxylate groups attached to the α-carbon participate in the formation of the peptide bond that links amino acids together to form a peptide or protein. The side chains do not participate in the formation of the covalent peptide bond and are thus available for bioconjugation. Whether a particular amino acid side chain on a VNP can serve as a target for functionalization depends on (i) the location on the viral capsid and (ii) the microenvironment of the side chain. Amino acids that are solvent-exposed on either the exterior or interior surface of the capsid are generally available for modification protocols. However, the reactivity of a specific residue also depends on the microenvironment, as some side chains participate in interactions with neighboring side chains and therefore may not be accessible for chemical bioconjugation. Common interactions include electrostatic interactions, hydrogen bonding, and, for Cys side chains, covalent disulfide bonds.

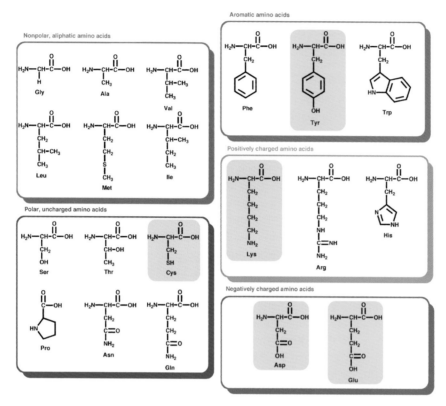

Figure 4.1 Overview of the 20 common amino acids. Amino acids used as targets for bioconjugation chemistry on VNPs are highlighted in yellow boxes. Gly, glycine; Ala, alanine; Val, valine; Leu, leucine; Met, methionine; Ile, isoleucine; Ser, serine; Thr, threonine; Cys, Cysteine; Pro, proline; Asn, asparagine; Gln, glutamine; Phe, phenylalanine; Tyr, tyrosine; Trp, tryptophan; Lys, lysine; Arg, arginine; His, histidine; Asp, aspartic acid; Glu, gutamic acid.

For bioconjugation, the most commonly utilized amino acids are Lys (amine-functional group), Asp and Glu (carboxylate-functional group), Cys (thiol-functional group), and Tyr (hydroxyl-functional group), all of which have been utilized as targets for modification of VNPs. Strategies targeting these amino acids will be discussed in this chapter. Other amino acids that can be utilized for bioconjugation chemistries are Arg, His, Met, and Trp; the reader is referred to the following textbooks for information on coupling techniques utilizing these amino acids: *Bioconjugate Techniques* (Hermanson, 1996) and *Bioconjugation* (Aslam & Dent, 1999).

4.1.1 Amine-Selective Chemistries

Amines are by far the most commonly used targets for bioconjugation chemistry. Primary amines in proteins are derived from the N-terminus (the terminal α-amine that is not engaged in a peptide bond) and the amine group of Lys side chains (Fig. 4.1), termed the ε-amine.

The ε-amine is a highly reactive nucleophile. A nucleophile is any atom containing an unshared pair of electrons or an excess of electrons able to participate in covalent bond formation. The reactivity stands in relation to the relative nucleophilicity and is dependent on whether the amine is protonated or unprotonated. An uncharged, unprotonated amine is more reactive compared with the protonated form. In its unprotonated form, the amine contains an unshared pair of electrons and is thus a good nucleophile. The Lys side chain is a strong base with a pK_a of 9.3–9.5 (pK_a is the acid dissociation constant and a measure of the strength of an acid in solution; the lower the value, the higher the extent of dissociation and strength of the acid). Based on the Henderson–Hasselbalch equation (Eq. 4.1), at a pH close to its pK_a 50% of the ε-amines are unprotonated and reactive. The realistic pK_a of a specific amino acid side chain in a protein is dependent on the microenvironment and is typically lower than the theoretical value; Lys modification can be carried out at pH values as low as pH 7.0 with negligible loss of reactivity, although of course this is dependent on the individual protein (Aslam & Dent, 1999; Hermanson, 1996).

$$pH = pK_a + \log\{[\text{base}]/[\text{acid}]\} \tag{4.1}$$

Amine-selective chemistries are a popular choice, and numerous reagents for amine modification are commercially available. For examples, browse the catalogs from Pierce (http://www.piercenet.com), Invitrogen (http://www.invitrogen.com), and Solulink (http://www.solulink.com). Reagents used for amine modification typically contain isothiocyanates or N-hydroxysuccinimide (NHS) esters as the functional group. Derivatization with carboxylate containing compounds can also be achieved using a carbodiimide as a coupling reagent (Aslam & Dent, 1999; Hermanson, 1996).

4.1.1.1 Isothiocyanate-mediated coupling

The coupling reaction of an isothiocyanate with an amine is outlined in Fig. 4.2 (panel A); coupling results in the formation of a stable thiourea bond. The most commonly used isothiocyanate-containing label is fluorescein–isothiocyanate (FITC), which has been conjugated to a large variety of VNPs.

4.1.1.2 *N*-Hydroxysuccinimide esters

Coupling of an NHS ester-activated compound to an amine results in the formation of a stable amide bond with the NHS as a leaving group (Fig. 4.2, panel B).

4.1.1.3 Carboxylates

Carboxylates can be covalently attached to amines via the formation of a peptide bond. To achieve chemical conjugation, the carboxylate-selective coupling agent 1-ethyl-3-(3-dimethylaminopropyl)carbodiimide (EDC) is typically used. EDC activates carboxylates by the formation of an *O*-acylisourea ester intermediate. This unstable intermediate reacts readily with primary amines, which results in the formation of a stable covalent amide bond (Fig. 4.2, panel C). NHS is often used in addition to EDC, where the NHS group reacts with the activated *O*-acylisourea and generates an NHS ester *in situ*, which readily reacts with primary amines (see above). This coupling technique can be applied toward coupling any molecule containing a carboxylate to a VNP with addressable Lys side chains.

4.1.1.4 Transamination

Covalent modification of the N-terminus can be achieved by site-specific transamination reactions mediated by the biological cofactor pyridoxal 5′-phosphate and has been applied to a genetically engineered version of the *Tobacco mosaic virus* (TMV) coat protein (Scheck *et al.*, 2008). In brief, the transamination allows the introduction of a ketone or aldehyde, both of which provide *orthogonal* ligation handles for further reaction such as oxime ligation (see Section 4.2.2). The word *orthogonal* comes from the Greek language and means "straight angle". In chemistry, *orthogonal* reactions are defined as strategies that allow the reaction of functional groups independently of each other. Orthogonal reactions are in general highly selective and reactive.

4.1.2 Carboxylate-Selective Chemistries

Carboxylate groups derived from the acids Asp and Glu (Fig. 4.1) can be used as targets for bioconjugation chemistry. The pK_a values of Asp and Glu side chains lie at around ≈4.5. The C-terminus of proteins (that is the terminal α-

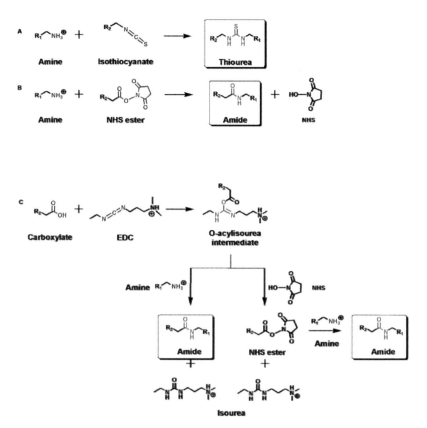

Figure 4.2 Reaction scheme for bioconjugation between primary amines and an isothiocyanate (A), a NHS ester (B), and a carboxylate in combination with the coupling reagents EDC and NHS (C). Figure provided by courtesy of Vu Hong (TSRI, La Jolla, CA, USA).

carboxylate that is not engaged in the formation of a peptide bond) can also be utilized; its pK_a is typically lower and lies at around ≈ 3.0–4.0. That means at physiological pH the carboxylates are ionized (unprotonated). However, even an unprotonated carboxylic acid, known as a carboxylate, is not a good nucleophile. In order to make use of carboxylates for bioconjugation, coupling reagents such as EDC and NHS are used (Aslam & Dent, 1999; Hermanson, 1996). The reaction is as outlined in Fig. 4.2 (panel C); however, in this case, the carboxylate is derived from the VNP and the amine from the molecule to be coupled.

Carboxylates derived from Asp and Glu have been exploited on a few VNPs (see Section 4.3.2); however, carboxylate-derivatization techniques are by far less popular compared with amine- and thiol-selective techniques.

4.1.3 Thiol-Selective Chemistries

Thiols derived from Cys side chains (Fig. 4.1) are highly reactive nucleo-philes. The theoretical pK_a of Cys lies at 8.8–9.1. At physiological pH, the Cys is unprotonated and reactive. Based on their highly reactive nature, Cys side chains in proteins are mostly found engaged in disulfide bonds. Cys side chains are typically not found in the free reactive form on the solvent-exposed surface of VNPs. There is only one example, where reactive and addressable Cys side chains have been reported: Cys side chains have been identified on the *interior* solvent-exposed surface of *Cowpea mosaic virus* (CPMV) (Wang *et al.*, 2002c, 2003b). Thiols are nevertheless useful groups to use for bioconjugation; therefore, a range of VNPs have been generated via genetic modification to add Cys side chains. Cys-added versions of *Cowpea chlorotic mottle virus* (CCMV), CPMV, *Flock House virus* (FHV), MS2, Ad, TMV, and M13, for example, are available (Destito *et al.*, 2009; Khalil *et al.*, 2007b; Klem *et al.*, 2003; Kreppel *et al.*, 2005; Miller *et al.*, 2007; Peabody, 2003; Wang *et al.*, 2002b).

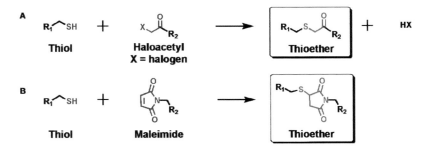

Figure 4.3 Reaction scheme for bioconjugation between a thiol and haloacetyl (A) and maleimide (B), respectively. Figure provided by courtesy of Vu Hong (TSRI, La Jolla, CA, USA).

A range of chemistries can be applied to thiols. Those that have been applied to VNPs include reaction with haloacetyl compounds (such as iodo- or bromoacetamides), coupling with maleimides, binding to gold, and use of small mercuric reagents. The reaction schemes for coupling haloacetyl- and maleimide-derivatives to thiols are given in Fig. 4.3; coupling results in the formation of a stable thioether bond (Aslam & Dent, 1999; Hermanson, 1996).

4.1.4 Tyr-Selective Chemistries

The aromatic group of Tyr side chains also provides a feasible target for protein modification. The phenyl group is, however, only modestly reactive, and to achieve bioconjugation the phenyl group must be converted into a phenolic side chain. The introduction of the hydroxyl group increases the nucleophilicity dramatically. The pK_a of the phenolic hydroxyl lies at 9.7–10.1 (Aslam & Dent, 1999; Hermanson, 1996).

Bioconjugation techniques utilizing Tyr side chains can be considered as more advanced. Commercial reagents are typically not available; thus, the starting materials and coupling reagents must be chemically synthesized. The most common reaction utilizing Tyr side chains on VNPs is diazonium coupling, which has been widely used on MS2 and TMV. One-electron oxidation and modification of Tyr side chains on CPMV have also been reported (see below).

4.1.4.1 Diazonium coupling

Diazonium coupling or azo coupling is the bioconjugation between an aniline (typically aromatic) and a Tyr side chain. The highly electron-deficient diazonium salt of the *p*-nitroaniline is widely used as a coupling reagent. Coupling results in the installation of an azo linkage *ortho* to the phenolic moiety of the Tyr side chain (Fig. 4.4). Francis and coworkers (University of California Berkeley, CA, USA) have made great progress in developing this coupling strategy for MS2 and TMV (Kovacs *et al.*, 2007; Schlick *et al.*, 2005). Introduction of a functional group to the VNP is achieved by using an aniline-containing derivative of the molecule of interest (Kovacs *et al.*, 2007). Functionalized aniline compounds are typically not commercially available and must be chemically synthesized. Alternatively, an alkyne-containing aniline can be attached; the alkyne can be regarded as a ligation handle for subsequent coupling reactions using click chemistry (Bruckman *et al.*, 2008) (see Section 4.1.2). Azo coupling has also been used to introduce aldehyde functionalities into the VNP scaffold; the aldehyde can serve as a target for oxime condensation reactions (Datta *et al.*, 2008; Hooker *et al.*, 2007, 2008; Kovacs *et al.*, 2007; Schlick *et al.*, 2005) (see Section 4.2.2).

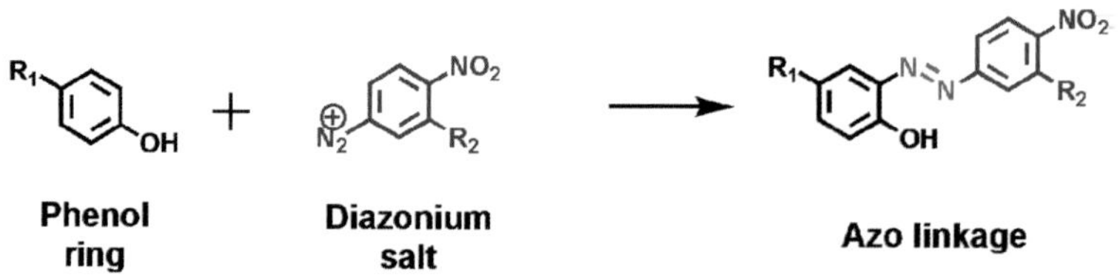

Phenol ring + **Diazonium salt** → **Azo linkage**

Figure 4.4 Bioconjugation of Tyr side chains using a diazonium salt of the *p*-nitroaniline; coupling results in a stable azo linkage. Figure provided by courtesy of Vu Hong (TSRI, La Jolla, CA, USA).

4.1.4.2 Photochemical oxidation of Tyr for cross-linking and labeling

Native CPMV particles display addressable Tyr residues. Structural data and the study of Tyr-minus mutants demonstrated that two Tyr side chains located in the S subunit are available for chemical conjugation (Meunier *et al.*, 2004). The Tyr phenol can be oxidized by one electron, allowing activation and subsequent bioconjugation. Two strategies were explored. In the first approach, Tyr residues were oxidized by one electron via treatment with the nickel(II) complex of the tripeptide Gly-Gly-His in the presence of magnesium monoperoxyphthalate (Ni/GGH/MMPP). In the second approach, the photochemical action of the tris(bipyridyl)ruthenium(II) dication was exploited. Both treatments mediated covalent conjugation of adjacent Tyr side chains and led to effective cross-linking of adjacent pentameric subunits. Besides cross-linking the Tyr side chain, functional groups can also be introduced using this type of chemistry. In a proof-of-concept study, fluorescein has been covalently attached to CPMV Tyr side chains (Meunier *et al.*, 2004).

4.1.5 Carbohydrate-Selective Chemistries

Some viruses, such as the archaeal virus *Sulfolobus islandicus* rod-shaped virus 2 (SIRV2), are glycosylated, allowing chemical modification using carbohydrate-selective chemistries (Steinmetz *et al.*, 2008a) (see also Section 4.3.5). In order to make use of carbohydrates, the hydroxyls must be converted into aldehydes. This is typically achieved under mild oxidation conditions using the reagent sodium *meta*-periodate. The aldehyde groups are reactive toward hydrazide conjugates and undergo a facile coupling reaction that results in the formation of a covalent hydrazone linkage (Fig. 4.5) (Aslam & Dent, 1999; Hermanson, 1996).

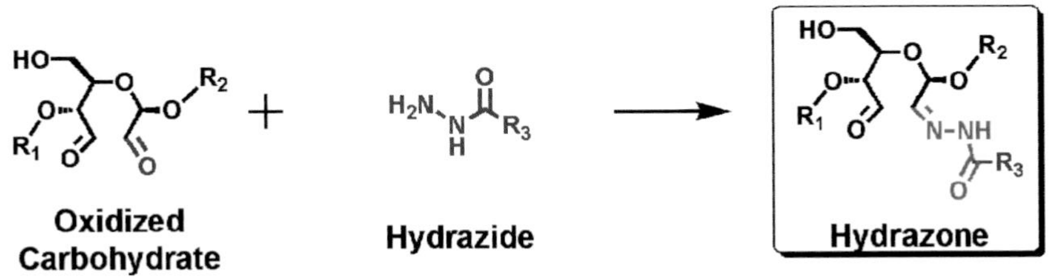

Oxidized Carbohydrate   **Hydrazide**   **Hydrazone**

Figure 4.5 Coupling reaction between an oxidized carbohydrate and a hydrazide derivative. Figure provided by courtesy of Vu Hong (TSRI, La Jolla, CA, USA).

4.1.5.1 Chemoselective glycoconjugation method

The feasibility of synthesis of *S*-linked glycoconjugates through site-specific ligation of 1-glycosyl thiols to proteins has been demonstrated utilizing the bacteriophage Qβ. The strategy exploits non-natural amino acid incorporation (recall Section 3.2.1), in this case L-homoallylglycine (L-Hag). Free-radical glycosylation reaction allows the synthesis of *S*-linked glycoconjugates (Fig. 4.6) (Floyd *et al.*, 2009). This method facilitates the incorporation of carbohydrates, which could be used for further modification. Alternatively, modified carbohydrates could be introduced.

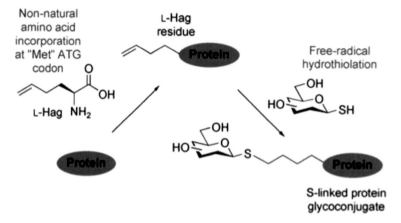

Figure 4.6 Synthesis of *S*-linked glycoconjugates through site-specific ligation of 1-glycosyl thiols to proteins. Reproduced with permission from Floyd, N., Vijayakrishnan, B., Koeppe, J. R., and Davis, B. G. (2009) Thiyl glycosylation of olefinic proteins: *S*-linked glycoconjugate synthesis, *Angew. Chem. Int. Ed. Engl.*, **48**(42), 7798–7802.

4.2 BIO-ORTHOGONAL REACTIONS: CLICK AND OXIME REACTIONS

There is need for alternative bioconjugation methods, such as the click reaction (Section 4.2.1) or oxime ligation (Section 4.2.2). Standard coupling procedures using NHS ester- or maleimide-activated reagents have slow reaction kinetics, and large excesses of reagents have to be used to facilitate efficient labeling. Click reaction and oxime ligation are highly efficient bioconjugation methods that require low concentration and excess of the reagent or ligand of interest. This is helpful when reagents are scarce or if solubility in aqueous conditions is a problem. Another advantage is lower costs, as less material is required.

4.2.1 Cu(I)-Catalyzed Azide–Alkyne Cycloaddition (CuAAC, a.k.a. Click Chemistry)

Bio-orthogonal reactions are those that involve functional groups that are inert to most biological molecules, and are thus of great interest for bioconjugation chemistry. The bio-orthogonal functional groups such as azide and alkyne have been very successfully applied. The reaction between an azide and an alkyne is catalyzed by Cu(I) and termed as Cu(I)-catalyzed azide–alkyne cycloaddition (CuAAC), also known as click chemistry. Click reactions have been widely used for bioconjugation and are a popular strategy because of their specificity, high yield, and wide range of solvents and pH stabilities (Kolb *et al.*, 2001).

The research team led by Finn (The Scripps Research Institute, La Jolla, CA, USA) has contributed enormously to the development of click reactions utilizing VNPs and virus-like particles (VLPs) as templates. To facilitate click conjugations on a VNP, an azide or alkyne functional group needs to be installed on the VNP scaffold. This can be achieved using a chemical or genetic approach. For most applications, the azide or alkyne functional group is covalently introduced by coupling a Lys-, Cys-, or Tyr-reactive azide or alkyne derivative to solvent-exposed amino acids on the particle surface (Fig. 4.7, panel A). Alternatively, azide- or alkyne-containing unnatural amino acids can be introduced into the capsid proteins under genetic control; this has been demonstrated using Qβ and *Hepatitis B virus* (HBV) VLPs (Strable *et al.*, 2008), as discussed in Chapter 3, Section 3.2.1.

The reaction between an azide and an alkyne is outlined in Fig. 4.7 (panel B). A copper-binding ligand is required to accelerate the reaction and prevent protein aggregation or degradation (Wang *et al.*, 2003a). The ligands, tris(triazolylmethyl)amine and sulfonated bathophenanthroline (Fig. 4.7, panel C) have been used for conjugation of a variety of molecules to VNPs. A new improved protocol has recently been established (Hong *et al.*, 2009). The new protocol offers the most convenient and efficient way to perform CuAAC by employing tris(3-hydroxypropyltriazolylmethyl) amine (THPTA), a water-soluble member of the tris(triazolylmethyl) amine family, which, in excess, intercepts reactive oxygen species generated in the coordination sphere of the metal. In addition, aminoguanidine is recommended to capture dehydroascorbate and its decomposition products before they can react with protein side chains. Click chemistry protocols have been established for CPMV, Qβ, *Potato virus X* (PVX), and TMV, and broad classes of molecules ranging from small chemical modifiers to intact proteins have been covalently attached (Bruckman *et al.*, 2008; Hong *et al.*, 2009; Kaltgrad *et al.*, 2008; Prasuhn *et al.*, 2007, 2008; Sen Gupta *et al.*, 2005a,b; Steinmetz *et al.*, 2009a,c; Udit *et al.*, 2008; Wang *et al.*, 2003a).

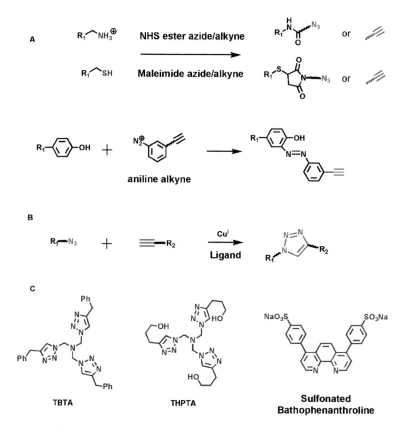

Figure 4.7 (A) Introduction of an azide or alkyne functional group by coupling and azide or alkyne derivative to an amine using a NHS ester, to a thiol using a maleimide, or phenol ring using an aniline derivative. (B) Reaction between an azide and alkyne resuls in the formation of a triazole linkage. (C) Copper-binding ligands used for click chemistry on VNPs: tris(triazoleylmethyl)amine (TBTA), THPTA, and sulfonated bathophenanthroline. Figure provided by courtesy of Vu Hong (TSRI, La Jolla, CA, USA).

4.2.2 Oxime Condensation

The bio-orthogonal reaction between an aldehyde and hydrazide or alkoxyamine is also highly chemoselective and efficient; this reaction is termed as oxime condensation. Oxime condensation reactions have been adapted for the VNPs MS2 and TMV (Datta *et al.*, 2008; Hooker *et al.*, 2007, 2008; Kovacs *et al.*, 2007; Schlick *et al.*, 2005). In order to perform this reaction, an aldehyde ligation handle is introduced into the VNP scaffold using either an aldehyde-containing NHS-derivative and coupling to Lys side chains, or an aldehyde-containing diazonium salt of *p*-aniline and coupling to Tyr side chains (Fig. 4.8, panel A). In the subsequent step, a hydrazide or alkoxyamine derivative of the compound of interest is added and covalently attached via the formation of a stable oxime bond (Fig. 4.8, panel B). For example, PEG chains and chelated gadolinium complexes have been attached to MS2 and TMV using this coupling strategy (Datta *et al.*, 2008; Hooker *et al.*, 2007; Kovacs *et al.*, 2007; Schlick *et al.*, 2005). A high-yielding and improved ligation approach has been developed that utilizes aniline as a catalyst to activate aromatic aldehydes toward amine nucleophiles (Dirksen & Dawson, 2008).

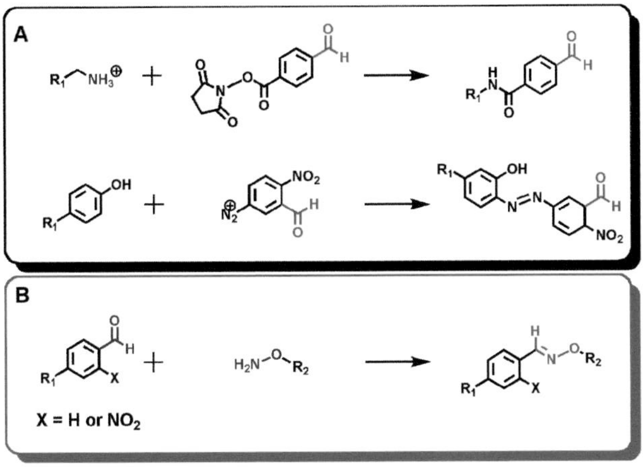

Figure 4.8 Oxime condensation reaction. (A) Attachment of an aldehyde to a Lys or Tyr side chain. (B) Oxime condensation between an aldehyde and alkoxyamine. Figure provided by courtesy of Vu Hong (TSRI, La Jolla, CA, USA).

4.3 EXAMPLES AND HIGHLIGHTS: ADDRESSABLE VNPs

A large variety of VNP building blocks have been utilized for bioconjugation strategies. Initial studies were concerned with the exploration of different coupling strategies. Standard chemistries using commercially available probes have been applied to nearly any VNP currently in use for nanotechnology. More advanced protocols facilitating the attachment and functionalization with nearly any chemical moiety within to intact proteins and other nanomaterials have also been developed. The following sections will give an overview of the addressable surface groups available on the different VNPs. Examples and highlights will be given; Table 4.1 at the end of this chapter will give an overview of the functional groups that have been attached to VNPs. This chapter will focus on the specifics of the chemical functionalization reaction. Applications of such modified hybrid VNP complexes are also given.

4.3.1 Solvent-Exposed Lys Side Chains on CPMV

Nearly all VNPs currently in use for nanotechnology offer solvent-exposed addressable Lys side chains. Functionalization and attachment is straightforward, and many reagents for Lys-modification are commercially available. For an overview of VNPs with addressable surface Lys residues, the reader is referred to Table 4.1 at the end of this chapter.

Bioconjugation has been extensively developed using the CPMV platform. CPMV particles have $P = 3$ symmetry and are assembled from 60 copies of each the S and L proteins. The S and L proteins form the so-called asymmetric unit, and 60 copies of the asymmetric unit form a virion (Lin *et al.*, 1999). In a proof-of-concept study, Lys reactivity was probed using fluorescein NHS ester and isothiocyanate derivatives (Fig. 4.9) Successful labeling was achieved under forcing conditions (long reaction time and large excess of labeling agents) where up to 240 labels could be attached (Wang *et al.*, 2002a,c).

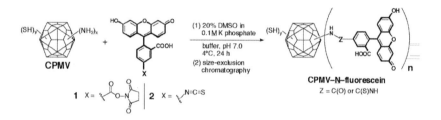

Figure 4.9 Attachment scheme for attachment of fluorescein NHS (1) and isothiocyanate (2) derivatives to surface Lys side chains on CPMV. Reproduced with permission from Wang, Q., Kaltgrad, E., Lin, T. Johnson, J.E., and Finn, M.G. (2002) Natural supramolecular building blocks: wild-type cowpea mosaic virus, *Chem. Biol.*, 9(7), 805–811.

In a follow-up study, the nature of the CPMV Lys-reactivity was addressed in greater detail. The X-ray structure and coordinates of CPMV [available at the Virus Particle ExploreR (VIPER) database at http://www.viper.scripps.edu] indicated the presence of five solvent-exposed Lys per asymmetric unit, which equates to 300 exposed Lys side chains per CPMV particle (Fig. 4.10) (Chatterji *et al.*, 2004a). To investigate whether each of these Lys side chains are available for bioconjugation, the Lys side chains were systematically deleted using side-directed mutagenesis. Lys side chains were changed into Arg side chains. The polarity and positive charge of the Arg side chains are likely to preserve the microenvironment of the original Lys residues, but are not expected to engage in the bioconjugation reactions. Unlike Lys, Arg side chains are poor nucleophiles. The theoretical pK_a of Arg side chain lies at around 12.0. At physiological pH (the pH at which the coupling reactions were conducted), the guanidine group is protonated and thus a poor nucleophile and not reactive toward NHS esters or isothiocyanates.

Single, double, triple, and quadruple Lys-minus mutants were generated and chemical labeling efficiency measured. The studies indicated that all of the five Lys are indeed available for bioconjugation, and the degree of labeling efficiency varies between the different sites. The most reactive groups were found to be Lys 38 on the S protein and Lys 99 on L (Chatterji *et al.*, 2004a).

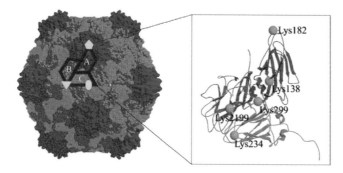

Figure 4.10 The structure of the viral capsid and the assymetric unit of CPMV. On the left is a space-filling model of CPMV capsid. The reference asymmetric unit is framed, and the symmetry elements are labeled. Small (S) subunits labeled A are in blue, and the large (L) subunits formed by two domains are in red (B domains) and in green (C domains). The oval represents a twofold axis, the triangle is a threefold axis, and the pentagon a fivefold axis. Shown on the right is a ribbon diagram of the asymmetric unit comprised of three jelly rolls — sandwiches, with surface Lys residues represented as spheres in cyan. Residue numbers are preceded by 1 if they are in the small subunit and 2 if they are in the large subunit. Lys138 (residue no. 38) and Lys182 (residue no. 82) are in the A domain, Lys299 and Lys234 are in the C domain, and Lys2199 is in the B domain. Reproduced with permission from Chatterji, A., Ochoa, W., Paine, M., Ratna, B. R., Johnson, J. E., and Lin, T. (2004) New addresses on an addressable virus nanoblock: uniquely reactive Lys residues on cowpea mosaic virus, *Chem. Biol.*, **11**(6), 855–863.

To demonstrate the level of spatial control and orientation that can be achieved using a VNP as a template for bioconjugation, NHS ester-activated nanogold was attached to either Lys 38 on S or Lys 99 on L using the respective mutants displaying a single reactive site. Structural analysis by cryo-electron microscopy and image reconstruction confirmed specificity of the reaction. Moreover, it was found that the different microenvironment impacted the presentation of the ligand (Fig. 4.11) (Chatterji *et al.*, 2004a). To date, this level of control has not been achieved with synthetic nanomaterials.

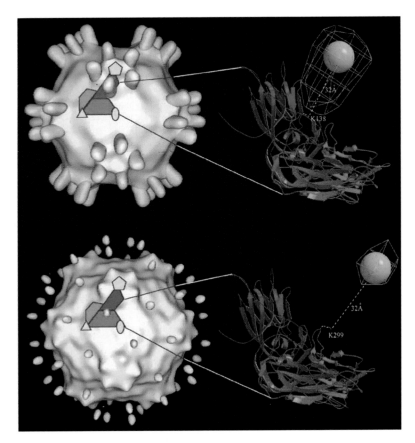

Figure 4.11 Images of gold-decorated CPMV mutants as determined by cryo-electron microscopy and image reconstruction. On top is shown a cryo-EM image of nanogold-labeled mutant vK138. The image is a composite of the native virus (gray) and difference map between the CPMV/gold conjugates and the native virus (gold). The asymmetric unit of the capsid and the symmetry operators are shown. The gold particles appear as spikes protruding from Lys138. Difference electron density, derived from vK138-gold conjugate, is superimposed with the ribbon diagram of the asymmetric unit of the virus capsid on the right. The steric constraint restricts the movement of gold particles. The A domain is represented in blue, B is shown in red, and C is denoted in green. The gold particle is drawn as a yellow sphere with a diameter of 14 Å. The center of the gold particle to the Lys residue is 32 Å. At the bottom is a cryo-EM image of nanogold/vK299 conjugate, similarly illustrated as for vK138 conjugate. The density corresponding to gold particles appears as islands, suggesting considerable latitudinal motion. As shown in the ribbon diagram, there is less constraint on the gold particles labeled at this site. Reproduced with permission from Chatterji, A., Ochoa, W., Paine, M., Ratna, B. R., Johnson, J. E., and Lin, T. (2004) New addresses on an addressable virus nanoblock: uniquely reactive Lys residues on cowpea mosaic virus, *Chem. Biol.*, **11**(6), 855–863.

4.3.2 Addressable Carboxylates from Asp and Glu Side Chains

Carboxylates derived from Asp and Glu are in general less favored for bioconjugation reactions. Compared with Lys, Cys, and Tyr side chains, carboxylates derived from Asp and Glu are less reactive. Chemical attachment of amine-containing compounds can be achieved making use of the coupling reagents EDC and NHS (see Section 4.1.2). An intrinsic problem with this strategy is that interparticle linkage can occur if the particle itself displays solvent-exposed amines; such is the case for most VNPs utilized (Table 4.1). Hydroxylamine can be added to stop the reaction and minimize the formation of VNP aggregates (hydroxylamines neutralize activated carboxylates) (Portney *et al.*, 2005).

Reactive carboxylates have been identified and utilized on the exterior surface of CPMV, CCMV, *Turnip yellow mosaic virus* (TYMV), M13, and SIRV2 (Bar *et al.*, 2008; Barnhill *et al.*, 2007a,b; Gillitzer *et al.*, 2002; Steinmetz *et al.*, 2008a,b). In addition, reactive carboxylates have been found on the interior surface of TMV and successful attachment of small chemical modifiers has been demonstrated (Schlick *et al.*, 2005) (see also Section 4.3.7 and Table 4.1).

4.3.3 Utilizing Cys Side Chains and the Role of Cys-Added Mutants

VNPs typically do not display any reactive Cys side chains on the solvent-exposed exterior capsid surface. In general, if Cys residues are present they are engaged in the formation of disulfide bonds. Based on the versatility of thiols for bioconjugation, a variety of Cys-added mutant VNPs have been generated and have been widely used in a variety of applications (see below). Naturally occurring reactive Cys side chains have been discovered and utilized on the solvent-exposed *interior* capsid surface of CPMV (discussed in Section 4.3.8).

Now, when designing a Cys-added mutant VNP, it is important to carefully choose the insertion site for the Cys residue. VNPs have a strong degree of symmetry and polyvalency, and a mutant particle will display multiple copies of the genetic modification on the particle surface. For example, a single mutation in a $T = 3$ particle will result in 180 modifications on the intact and assembled virion. Based on the reactivity of the thiol group of Cys side chains, an undesired effect often observed with Cys-added mutants is that the particles heavily aggregate via the formation of disulfide bonds between adjacent particles. Therefore, it is critical to carefully design Cys-added mutants.

In the case of CPMV, the first attempt to generating a CPMV Cys-added mutant exploited insertion of the Cys residues in the highly solvent-exposed βB–βC loop at amino acid position 25 of the S protein (Wang *et al.*, 2002b) (Fig. 4.12). The mutants were found highly reactive toward thiol-selective chemistries. On the downside, the particles show strong tendency to form interparticle aggregates via the formation of disulfide bonds in the absence of high concentrations of reducing agents. After 2–3 months of storage, the particles undergo such extensive interparticle linkage that the virus solution can turn into a viscous "glue" (Fig. 4.12) (Steinmetz *et al.*, 2007), thus limiting the usefulness of this particular mutant.

Figure 4.12 Left panel: location of the Cys residue in the solvent-exposed βB–βC loop at amino acid position 25 in the CPMV capsid. CPMV consists of small (S, in blue) and large (L, in green) subunits. The surface-exposed βB–βC loop (highlighted in yellow) lies on the S subunit and it is composed of the following nine amino acids: Thr, Pro, Pro, Ala, Pro, Glu, Ser, Asp, and Val. The CPMV Cys-added mutant contains a point mutation: exchange of a Ser at amino acid position 25 on the S subunit with a Cys residue (highlighted as a sphere). Right panel: aggregation of mutant CPMV particles displaying a cysteine residue in the solvent-exposed βB–βC loop.

Efforts have been made toward "smart" mutant design, and a large collection of Cys-added CPMV mutants has been evaluated (reviewed in Strable & Finn, 2009). As a general design rule, the Cys residue should be inserted on the solvent-exposed surface at a site that is somewhat buried on the particle surface. This facilitates reactivity toward thiol-selective chemistries, but avoids undesired aggregation effects.

Cys-added versions of CCMV, CPMV, FHV, MS2, Ad, TMV, and M13 (Destito *et al.*, 2009; Khalil *et al.*, 2007a; Klem *et al.*, 2003; Kreppel *et al.*, 2005; Miller *et al.*, 2007; Peabody, 2003; Wang *et al.*, 2002b) have been made and utilized as building blocks for the construction of nanocircuits (Blum *et al.*,

2005a) and as tracers for biosensors (Sapsford *et al.*, 2006; Soto *et al.*, 2006), and have been developed for light-harvesting systems (Endo *et al.*, 2007; Ma *et al.*, 2008; Miller *et al.*, 2007) (as summarized in Table 4.1 and Section 4.4).

There is high demand for generating multifunctional devices, that is a VNP comprising different functionalities. For example, for the development of "smart" devices for drug delivery, one has to combine at least two functional groups into a single formulation, that is a targeting ligand (to give tissue specificity) and a therapeutic molecule (to treat the diseased tissue). Generation of Cys-added mutants is of importance; with the introduction of the Cys residues, not only additional attachment sites are introduced but also a different type of target, thus allowing sequential labeling of different sites exploiting different site-selective chemistries.

4.3.3.1 Introducing thiols—a chemical approach

Thiols are highly reactive toward a broad collection of chemicals and thus are useful for bioconjugation. As Cys mutants of VNPs can be challenging to work with and "smart" design can be time-consuming, alternative strategies have been employed allowing the introduction of reactive thiols by means of chemistry. Chemically, engineered thiolated VNPs can be generated by making use of the coupling reagent *N*-succinimidyl-*S*-acetylpropionate (SATP), a compound containing a protected acetylated thiol. Covalent attachment is achieved via coupling Lys side chains. To facilitate thiol-selective chemistry, the protecting group can be removed by deacetylation. This method was found to be a useful addition to generating Cys mutants. Thiolated VNPs were found to be stable, did not aggregate over a monitored time frame of 3 months, and were reactive toward maleimide-specific chemistries (Steinmetz *et al.*, 2007).

4.3.4 Spatially Controlled Labeling of Rod-Shaped VNPs

Rod-shaped particles can potentially be labeled with sequential and spatial control at the ends versus the virus body. End-specific labeling has been demonstrated for the rods SIRV2, TMV, and M13 (Nam *et al.*, 2004; Steinmetz *et al.*, 2008a; Yi *et al.*, 2005, 2007). End labeling of M13 is discussed in Section 4.3.8.

The capsids of the rod-shaped archaeal virus SIRV2 are composed of two different coat proteins, a major coat protein that forms the virus body and a minor coat protein that forms the end structures of the virions. It has been shown that different chemistries can be applied to the different

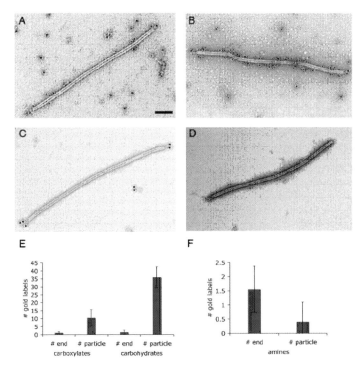

Figure 4.13 Immunogold staining of biotinylated SIRV2 particles using gold-labeled anti-biotin antibodies and transmission electron microscopy. (A) Biotinylated SIRV2 particles labeled using carboxylate-selective chemistry, (B) biotinylated SIRV2 particles labeled at carbohydrates, (C) biotinylated SIRV2 particles labeled using amine-selective bioconjugation techniques, and (D) non-modified SIRV2 used as a control. (E, F) Ten particles each were further analyzed and the gold labels per particle at the virus body and the virus end structures (tail fibers), respectively, counted. Note difference in scale of (E) and (F). (F) Statistical analysis of differences between virus body and end labeling was performed using Student's two-tailed t-test (Microsoft Excel) where $P = 0.0078$. Reproduced with permission from Steinmetz, N. F., Bize, A., Findlay, K. C., Lomonossoff, G. P., Manchester M., Evans, D. J., and Prangishvili, D. (2008) Site-specific and spatially controlled addressability of a new viral nanobuilding block: sulfolobus islandicus rod-shaped virus 2, *Adv. Funct. Mater.*, **18**, 3478–3486.

structural proteins. Both the major and minor coat proteins are accessible for modification using carboxylate- and carbohydrate-selective chemistries. However, only the minor coat protein that forms the end structure can be labeled using amine-selective chemistries (Fig. 4.13) (Steinmetz *et al.*, 2008a).

The ability to selectively attach labels and functionalities to the end structures versus virus body opens the door for the development of highly specialized and multifunctional devices. Many applications in nanotechnology deal with the immobilization of nanomaterials on solid devices. The bifunctionality of rod-shaped particles can be exploited to utilize the end structures for selective binding onto surfaces. This has been demonstrated using TMV. Researchers have made use of the polar nature of TMV; the helical encapsulation of the RNA molecule results in sequence-definable 5′ and 3′ ends. A mild disassembly protocol was used to partially disassemble the protein coat and expose the RNA at the 5′ end. Spatially oriented assembly of TMV on solid supports was achieved in a controlled manner via nucleic acid hybridization using complementary oligonucleotides (Yi *et al.*, 2005, 2007) (see also Chapter 7).

4.3.5 Breaking the Symmetry of Icosahedral Particles

Icosahedral VNPs are highly polyvalent and symmetrical particles. For many applications, polyvalency and high degree of symmetry are desired. However, for some applications it may be desired to break the symmetry of the VNPs. For example, it can be challenging to control immobilization of VNPs directionally and orientationally because of multivalency. A range of Cys-added mutant VNPs have been utilized for immobilization of gold surfaces (see also Chapter 7). Because of the multivalent display of Cys residues on the exterior surface and the propensity of thiols to undergo formation of disulfide linkages, binding of Cys-added VNPs onto solid supports can result in the formation of interparticle aggregates. Efforts to break the symmetry of VNPs have been made utilizing the platform CCMV. A solid-state synthetic approach has been utilized to break the symmetry of Cys-containing CCMV mutants (Fig. 4.14). The method utilizes a three-step protocol: in step 1, the Cys-mutant of CCMV was immobilized on a Sepharose resin containing an activated thiol. Unbound particles were removed by extensive washing. In step 2, all remaining free thiols on the virus surface were chemically passivated using iodoacetic acid (IAA) that readily reacts with free thiols. The passivated CCMV particles were then eluted from the resin in step 3 by treatment with 2-mercaptoethanol, a chemical that reduces the disulfide bonds that are holding the VNPs attached to the resin. After elution, only the Cys side chains that were engaged in binding to the resin are free and remain reactive. Immobilization of symmetry-broken particles was tested in comparison with wild-type particles (that do not contain any solvent-exposed Cys side chains), passivated mutant particles (in which all the Cys side chains had been passivated), and non-treated Cys-added mutant particles (display 180 free Cys side chains). Wild-type and passivated mutant particles did not bind onto gold surfaces. Non-treated Cys mutants formed extensive aggregates. Only the symmetry-broken particles facilitated formation of a controlled monolayer on the substrate (Fig. 4.14) (Klem *et al.*, 2003).

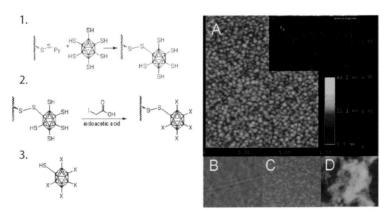

Figure 4.14 Symmetry breaking of a Cys-added mutant CCMV particle. Left panel: outline of the solid-state syntetic approach. Step 1: binding of the particles to the thiol-activated resin. Step 2: passivation of unbound cysteine residues with IAA. Step 3: elution of the symmetry-broken particles by reduction. Right panel: tapping-mode atomic force microscopy imaging on CCMV particle formulations added onto a gold substrate: (A) symmetry-broken particles with distance and height profile (inset); (B) wild-type CCMV with no exposed thiols; (C) Cys-mutant treated with IAA to passivate all exposed thiols; (D) untreated Cys mutant particles with 180 surface-exposed thiols. All scans shown are 2 µm in length. Reproduced with permission from Klem, M. T., Willits, D., Young, M., and Douglas, T. (2003) 2-D array formation of genetically engineered viral cages on Au surfaces and imaging by atomic force microscopy, *J. Am. Chem. Soc.*, **125**(36), 10806–10807.

In a different approach, breaking the symmetry was achieved by making use of *in vitro* dis- and re-assembly techniques; controlled and sequential ligand display was facilitated through mixed self-assembly (Fig. 4.15) (Gillitzer *et al.*, 2006). Herein, CCMV capsids were independently decorated with two different types of ligands to generate two populations of labeled virions: type I labeled with ligand A (biotin) and type II labeled with ligand B (digoxigenin). The particles were then *in vitro* disassembled and the resulting coat protein subunits separately purified. Re-assembly was performed using controlled ratios of type I and II subunits, exerting control of the stoichiometry of ligands A and B displayed on the final assembled virions (Fig. 4.15) (Gillitzer *et al.*, 2006).

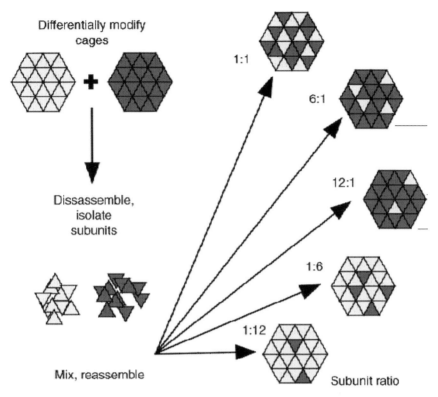

Figure 4.15 Schematic for the assembly of multifunctionalized CCMV particles. Two populations of particles are labeled and disassembled, and their subunits are purified. The differentially labeled subunits are subsequently mixed together at different ratios during re-assembly, resulting in multifunctional particles. Reproduced with permission from Gillitzer, E., Suci, P., Young, M., and Douglas, T. (2006) Controlled ligand display on a symmetrical protein-cage architecture through mixed assembly, *Small*, **2**(8–9), 962–966.

4.3.6 Attachment of Proteins and Nanomaterials

Coupling strategies also allow the attachment of larger complexes such as intact proteins, antibodies, or nanomaterials including quantum dots (QDs), single-walled carbon nanotubes (SWCNTs), and fullerenes to VNPs (Chatterji *et al.*, 2004b; Portney *et al.*, 2005; Sapsford *et al.*, 2006; Sen Gupta *et al.*, 2005a; Soto *et al.*, 2006; Steinmetz *et al.*, 2009b; Suci *et al.*, 2007a). This can be achieved using homo- and heterobivalent chemical linkers and covalent coupling; non-covalent strategies have also been explored (see Section 4.3.8).

4.3.7 Labeling the Interior Capsid Surface Via Bioconjugation

Besides utilizing the exterior solvent-exposed surface of VNPs, the interior surface can also be utilized for bioconjugation. This has been shown for CPMV, MS2, and TMV (Endo *et al.*, 2007; Hooker *et al.*, 2004, 2007; Ma *et al.*, 2008; Miller *et al.*, 2007; Schlick *et al.*, 2005; Wang *et al.*, 2002b,c).

Small chemical modifiers can, in general, diffuse freely between the bulk medium and the interior cavity of VNPs, and thus be attached to available targets on the interior capsid surface. For example, interior surface modifications can be achieved using CPMV, which displays reactive Cys side chains on the interior capsid surface. As CPMV does not display any Cys residues on the exterior surface, thiol-selective chemistries are specific for the interior (Wang *et al.*, 2002b,c).

Modification of the interior surface has been extensively studied and utilized using the bacteriophage MS2 (Hooker *et al.*, 2004, 2007; Kovacs *et al.*, 2007). Modification of the interior cavity of MS2 is more feasible compared with modification of the interior of CPMV. The particles of CPMV have 12 pores located at the fivefold axis with a diameter of about 0.7 nm (Lin *et al.*, 1999). MS2 offers 32 pores at the five- and threefold axes with a diameter of 1.8 nm (Golmohammadi *et al.*, 1996), making the interior more accessible. MS2 offers a highly reactive Tyr residue on the interior capsid surface that has been utilized for various bioconjugation strategies, and a selection of imaging molecules including fluorescent dyes and gadolinium complexes used in MRI have been successfully introduced (Hooker *et al.*, 2004, 2007; Kovacs *et al.*, 2007). The possibility of modifying the interior of VNPs in addition to attaching functionalities to the exterior surface is beneficial, especially when developing multifunctional devices. For example, probes for tissue-specific imaging can be designed in such a way that the imaging modality is inside the VNP and tissue-specific targeting ligands could be attached to the exterior surface (discussed in detail in Chapter 8).

The interior of the rod TMV can also be utilized; small organic dyes can be covalently attached at two interior Glu residues (Schlick *et al.*, 2005). Besides using intact TMV particles, interior labeling can also be achieved by disassembling the particles into coat protein monomers, followed by labeling and reassembly. This technique has been used for the development of light-harvesting systems (Endo *et al.*, 2007; Ma *et al.*, 2008; Miller *et al.*, 2007) (Section 4.4.3).

4.3.8 Non-Covalent Coupling Methods

Biospecific interactions, such as biotin-streptavidin or hexa-His-nickel–nitrilotriacetic acid (Ni-NTA) have also been used for interlinking VNPs with novel functionalities. These interactions have also been widely used for immobilization of VNPs on solid supports and the fabrication of multilayered arrays, which will be discussed in Chapter 7.

Biotinylated CCMV particles were joined with biotinylated antibodies using streptavidin as a linker (Fig. 4.16). Antibody-interlinked CCMV particles were then utilized for targeting biofilm-forming bacteria. These principles were extended, and dual-functionalized targeted, therapeutic CCMV particles were designed. Ruthenium complexes were attached to antibody-interlinked CCMV particles; the ruthenium complexes served as photosensitizers for photodynamic cell killing (Suci *et al.*, 2007a,b) (see also Chapter 8).

A range of non-covalent strategies has been developed to specifically link targeting molecules to the surface of Ad vectors to achieve tissue-specific targeting of the gene delivery vectors (reviewed in Barnett *et al.*, 2002; Campos & Barry, 2007; Mizuguchi & Hayakawa, 2004; Singh & Kostarelos, 2009). This includes the use of fusion proteins that combine a capsid-binding domain and a cell-binding ligand. Antibodies or antibody fragments specific for the Ad coat proteins or soluble coxsackievirus and adenovirus receptor (CAR) domain (which is exposed on the Ad particle surface) are typically used as capsid-binding domains. Cell- and tissue-specific ligands can be peptides, proteins, antibodies, or antibody fragments. Some examples include epidermal growth factor (EGF), fibroblast growth factor (FGF-2), gastrin-releasing peptides (GRPs), and antibodies targeted to carcinoembryonic antigen (CEA). Similar strategies have been explored for the gene delivery vehicle AAV (reviewed in Baker, 2003; Kwon & Schaffer, 2008).

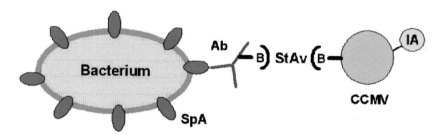

Figure 4.16 Cartoon showing the strategy used to target CCMV to *Staphylococcus aureus* (bacterium); streptavidin (StAv) was used to couple biotinylated anti-SpA antibody (Ab) to CCMV, which was dual labeled with biotin (B) and an imaging agent (IA). Reproduced with permission from Suci, P. A., Berglund, D. L., Liepold, L., Brumfield, S., Pitts, B., Davison, W., Oltrogge, L., Hoyt, K. O., Codd, S., Stewart, P. S., Young, and M., Douglas, T. (2007) High-density targeting of a viral multifunctional nanoplatform to a pathogenic, biofilm-forming bacterium, *Chem. Biol.*, **14**(4), 387–398.

The utility of mutant VNPs displaying hexa-His tags as templates for functionalization has also been shown. Mutant particles of P22, Ad, and CPMV were successfully labeled with nanogold particles using Ni-NTA linkers (Chatterji *et al.*, 2005; Kang *et al.*, 2008; Saini *et al.*, 2008). Besides introducing functionalities using the His affinity tag, one can also make use of the tag to achieve immobilization of the VNPs. This has been shown for Qβ and CPMV (Medintz *et al.*, 2005; Udit *et al.*, 2008).

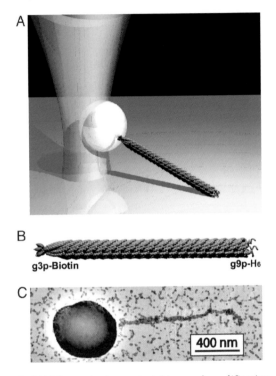

Figure 4.17 Single M13 bacteriophage stretching and modification. (A) Rendering of M13 bacteriophage stretching by an optical trap (not to scale). (B) The end complexes of M13 (identified in purple) consist of several copies of genetically modifiable pIX and pIII monomers, which have been modified with hexa-His tags and biotin, respectively. (C) A transmission electron microscopic image of the assembly architecture, where a gold-binding peptide was fused to pVIII monomers (forms the virus body) for gold nanoparticle incorporation along the length of the capsid. Reproduced with permission from Khalil, A. S., Ferrer, J. M., Brau, R. R., Kottmann, S. T., Noren, C. J., Lang, M. J., and Belcher, A. M. (2007) Single M13 bacteriophage tethering and stretching, *Proc. Natl. Acad. Sci. USA*, **104**(12), 4892–4897.

The biotin–streptavidin and hexa-His-Ni-NTA system can also be combined on a single VNP. M13 particles, which display different proteins at each end of the particles, were genetically modified with different tags at each end: a streptavidin-binding peptide was inserted as pIII fusion and a hexa-His tag as pIX fusion, respectively (see Fig. 4.17, panel B for the location of pIII and pIX on the M13 virion). The addition of a heterobifunctional linker—a streptavidin–Ni-NTA complex—to the modified phages resulted in reversible ring formation of the flexible filamentous particle (Nam *et al.*, 2004) (see also Chapter 7).

In a similar approach, M13 was modified with a hexa-His tag as a pIX fusion. At the other end of the phage, selenocysteine (thiol-functional group) was genetically introduced in pIII; a biotin functionality was subsequently introduced using thiol-selective chemical coupling. The dual-functionalized phages were then used as biomolecular tethers, and the elasticity of the biopolymer was tested using an atomic force microscopy (AFM) approach. In brief, the phage was immobilized on a surface via the hexa-His tag (pIX); the other end (pIII) was attached to an AFM tip using a streptavidin-linker protein (Fig. 4.17) (Khalil *et al.*, 2007a).

4.3.9 General Coupling Protocols, Purification, and Characterization Methods

Coupling procedures and reaction conditions vary widely, and optimized protocols have been developed. The reader is referred to the references given throughout this chapter and Table 4.1.

Once chemically modified, the functionalized VNP has to be purified from the reactants and coupling reagents. This is typically achieved using density gradient ultracentrifugation (Fig. 4.18) followed by ultrapelleting. With applications in medicine and materials emerging, quality control and verification of the integrity of the particles after modification are important. The integrity of VNPs can be verified using a range of methods, such as electron microscopy, size exclusion chromatography, ion exchange chromatography, dynamic light scattering, UV/visible spectroscopy, and native and denaturing gel electrophoresis. It is difficult to generalize techniques used for testing whether the chemical functionality was successfully introduced and to quantify the labels per particle. A combination of the above-mentioned techniques will give a good idea of whether a VNP has been successfully modified; for example, if modified with a fluorescent molecule, the proteins can be visualized under UV light in density gradient (Fig. 4.18), and in native or denaturing gels. UV/visible spectroscopy can be used to quantify the number of labels. However, additional and more specific techniques may be chosen for the characterization of the hybrid VNP material, which depend on the biophysical and biochemical properties of the label itself.

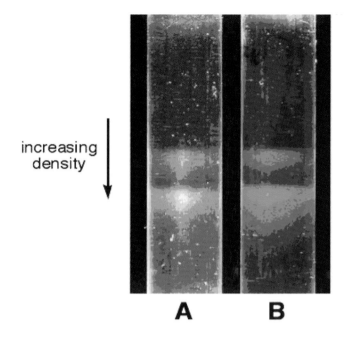

increasing
density

A   B

Figure 4.18 Analysis of dye-labeled VNPs. Ultracentrifugation of native (A) and fluorescein-labeled CPMV particles (B) through sucrose gradients. The two bands in each sample contain virus particles encapsulating the two different RNA molecules, RNA-1 and RNA-2, of the genome and are referred to as bottom and middle components, respectively. Top components, which are devoid of RNA, only make a small amount of most CPMV preparations and are not visible here. Reproduced with permission from Wang, Q., Kaltgrad, E., Lin, T., Johnson, J. E., and Finn, M. G. (2002) Natural supramolecular building blocks: wild-type cowpea mosaic virus, *Chem. Biol.*, **9**(7), 805–811.

4.4 APPLICATIONS OF CHEMICALLY LABELED VNPs

The VNP serves, in many cases, as a template or scaffold for functionalization. It can be regarded as a platform for presentation of functional ligands. VNPs comprise many copies of identical coat protein subunits and are thus highly polyvalent. Polyvalent display is desired for a range of applications, especially for applications in medicine (discussed in Chapter 8). Multivalent display is also beneficial for the development of novel sensors; multivalent display of signaling molecules leads to signal enhancement. VNPs have been combined with redox-active molecules to yield novel electro-active materials for potential applications in sensors, catalytic devices, or electronic devices. Dye-labeled VNPs sensors have been utilized in biological assays for improved detection sensitivity. Last but not least, light-harvesting systems were fabricated making use of fluorophore-labeled coat proteins and self-assembly methods. Some examples are discussed below (see also Table 4.1).

4.4.1 Electro-Active CPMV Complexes

To test the potential of VNPs to serve as a platform for the development of redox-active materials, CPMV particles were covalently modified with redox-active species. Decoration of CPMV with approximately 240 ferrocenes and 180 viologen moieties has been achieved (Steinmetz *et al.*, 2006b,c). In both cases, electrochemical studies confirmed the presence of redox-active nanoparticles; the redox couple was electrochemically reversible. Simultaneous multielectron transfer was observed, indicating that the multiple redox centers are independent and behave as essentially non-interacting redox units. The resulting robust and monodisperse particles could serve as multielectron reservoirs and nanoscale electron transfer mediators in redox catalysis or amperometric biosensors.

4.4.1.1 Self-assembly of conducting networks on the surface of CPMV

CPMV particles have been used as a scaffold for a bottom-up self-assembly approach to generate conductive networks at the nanoscale (Blum *et al.*, 2005b). This study highlights how the use of VNPs facilitates high-precision control over the attachment sites. In brief, two different CPMV Cys-added mutants were designed that displayed the Cys-reactive residues in different surface loops on the capsid surface. One mutant displayed 60 Cys residues, whereas a double mutant offered 120 Cys-derived attachment sites. Gold nanoparticles were bound to the genetically engineered sites based on gold–sulfur interaction (Fig. 4.19) (Blum *et al.*, 2005b). The attached gold nanoparticles were subsequently interconnected by molecular wires, thus creating a 3D conducting molecular network. To achieve this, the following molecules were used: $1,4\text{-}C_6H_4[trans\text{-}(_4\text{-}AcSC_6H_4C\equiv CPt\text{-}(PBu_3)_2C\equiv C]_2$ and oligophenylenevinylene. Molecular attachment of the linkers was confirmed by fluorescence spectroscopy. Physical measurements confirmed the conductance of the self-assembled molecular network (Blum *et al.*, 2005b).

Not only can materials be attached with high precision to the VNP surface, but also materials can be interlinked on the particle surface. This holds great promise in assembling and interconnecting conducting materials on the nanometer scale and may lead to the development of nanoelectronic systems, such as nanocircuits and data storage devices.

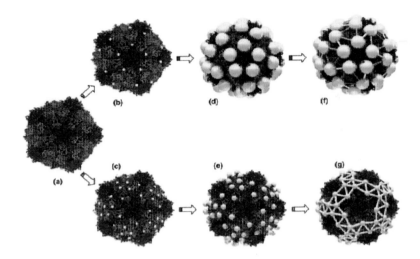

Figure 4.19 Schematic of the procedure used to create molecular networks on the surface of the virus capsid: (A) CPMV capsid structure from crystallographic data; (B) mutant with one cysteine (white dots) per subunit. The four nearest-neighbor cysteine-to-cysteine distances are 5.3, 6.6, 7.5, and 7.9 nm; (C) double mutant with two cysteines per subunit. The four nearest-neighbor cysteine-to-cysteine distances are 3.2, 4.0, 4.0, and 4.2 nm; (D) mutant with 5-nm gold nanoparticles bound to the inserted cysteines; (E) double mutant with 2-nm gold nanoparticles bound to the inserted cysteines; (F) mutant with the 5-nm gold particles interconnected using 1,4-C$_6$H$_4$[*trans*-($_4$-AcSC$_6$H$_4$C≡CPt-(PBu$_3$)$_2$C≡C]$_2$ (red) and oligophenylenevinylene (silver) molecules; (G) double mutant with the 2-nm gold particles interconnected with oligophenylenevinylene molecules. Reproduced with permission from Blum, A. S., Soto, C. M., Wilson, C. D., Brower, T. L., Pollack, S. K., Schull, T. L., Chatterji, A., Lin, T., Johnson, J. E., Amsinck C., Franzon P, Shashidhar, R., and Ratna, B. R., (2005) An engineered virus as a scaffold for three-dimensional self-assembly on the nanoscale, *Small*, **1**(7), 702–706.

4.4.2 Fluorescent-Labeled VNPs as Signal Enhancers in Biological Assays

Biological assays such as DNA microchip arrays and immunoassays make use of fluorescent-labeled molecules for detection. Because of the multivalent nature of VNPs, multiple fluorescent dyes can be displayed on a single VNP, thus enhancing signal sensitivity.

In a first example, VNP sensors were utilized in a DNA microarray used for pathogen genotyping. Herein, CPMV particles were covalently modified with approximately 40 dyes. Neutravidin (a protein that binds biotin) was also attached to the VNPs and served as a recognition element.

The VNP sensors were then evaluated in a DNA microarray as outlined in Fig. 4.20 (Soto *et al.*, 2006). The target DNA on the chip is probed with biotinylated test DNA. If the sequences are complementary (i.e., the nucleotide sequences anneal to each other by virtue of base pairing), the test DNA binds to the target DNA via nucleic acid hydridization. Next, the dual-functionalized VNPs are added; binding is accomplished via interaction of the conjugated neutravidin and the biotinylated test DNA. Detection is carried out via the fluorescent signal derived from the fluorophores attached to the VNP. Signal sensitivity of CPMV sensors was compared to the signal intensity given when using: (i) dye-labeled test DNA molecules (this is one-step detection method; the dye-labeled DNA is used instead of biotinylated DNA) and (ii) dye-labeled streptavidin. The CPMV sensors outperformed the conventionally used detection devices (Soto *et al.*, 2006).

In a second example, dye- and antibody-conjugated CPMV probes were utilized in an immunoassay. Again high sensitivity and specificity were confirmed (Sapsford *et al.*, 2006). These techniques, of course, have the potential to be applied to other VNPs as well.

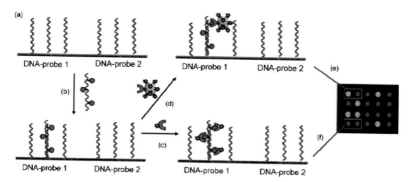

Figure 4.20 DNA microarray detection scheme. (A) DNA oligonucleotides 1 and 2 (probes) are immobilized in a microarray format on glass slides. (B) DNA probe 1 hybridizes with a previously amplified and biotinylated target DNA molecule. This hybridization event is detected using (C) streptavidin–Cy5 (Cy5 = fluorescent dye) or (D) NA–Cy5–CPMV (NA = neutravidin). (E) Quantification post-detection indicates a true-positive signal for the NA–Cy5–CPMV detection method (blue spot, hybridization with DNA probe 1) or a true-negative signal (gray spot, non-hybridization with DNA probe 2). (F) A false-negative signal (gray spot) is observed for streptavidin–Cy5 as the total number of fluorophores at the DNA probe 1 spot results in the generation of a signal that is below the detection threshold (or background). Reproduced with permission from Soto, C. M., Blum, A. S., Vora, G. J., Lebedev, N., Meador, C. E., Won, A. P., Chatterji, A., Johnson, J. E., and Ratna, B. R. (2006) Fluorescent signal amplification of carbocyanine dyes using engineered viral nanoparticles, *J. Am. Chem. Soc.*, **128**(15), 5184–5189.

4.4.3 Light-Harvesting Systems

Light-harvesting systems in nature allow the conversion of light into energy (Nelson & Ben-Shem, 2004). In plants, this process is called *photosynthesis*. The energy of sunlight is harvested and converted into glucose (see Eq. 4.2).

$$\text{Sunlight} + 6H_2O + 6CO_2 \rightarrow C_6H_{12}O_6 \text{ (glucose)} + 6O_2 \tag{4.2}$$

In plants, the green pigments (chromophores) called chlorophylls collect energy from light. Chromophores transport energy through a mechanism referred to as *Forster resonance energy transfer* (FRET). When a photon is absorbed by a chromophore, it causes electron excitation, which means the excited electron is raised to a higher energy level. The electron eventually returns to the non-activated lower energy level and light is emitted. In FRET, two chromophores are coupled, and the excitation energy is transferred from one chromophore to another, from a donor to an acceptor, that is when the energy levels of absorption and emission match. The distance between the chromophores plays an important role. Nature has developed precisely spaced arrays of chromophores that facilitate light harvesting and conversion into energy (Nelson & Ben-Shem, 2004).

Artificial light-harvesting systems are of great interest as they could potentially be used in solar cells and allow the conversion of sunlight into electrical energy. Protein scaffolds, such as the coat proteins of VNPs allow the precise positioning of chromophores through chemical bioconjugation. A range of systems based on the TMV platform, all based on similar design principles, have been developed; here we will highlight one of these studies (Endo *et al.*, 2006, 2007; Ma *et al.*, 2008; Miller *et al.*, 2007).

TMV coat proteins were modified with different chromophores that function as donor and acceptor sites, and then assembled into disks and rods. The chromophores are brought in proximity through self-assembly of the coat proteins and enable FRET to occur. In brief, TMV coat protein monomers with genetically introduced Cys residues were modified with three different chromophores through maleimide coupling. Oregon Green served as the primary donor, rhodamine was used as an intermediate donor, and Alexa Fluor 594 was used as the acceptor. The different labeled coat proteins were then self-assembled into disk and rod structures making use of the well-understood *in vitro* self-assembly mechanism of TMV. TMV rod assemblies contain 700 chromophores per 100-nm rod; the spacing between the chromophores lies within the distance range in which FRET occurs. Horizontal spacing was 1.8 nm and vertical spacing 2.3 nm. Rod-assemblies with varying donor:acceptor were assembled and evaluated using fluorescence spectroscopy. A complex system containing the three different chromophores was assembled with a final ratio of Oregon Green: rhodamine:Alexa Fluor 594 of 8 : 4 : 1 and tested. Efficient light transfer and light harvesting were confirmed (Fig. 4.21) (Miller *et al.*, 2007). Such systems are regarded as promising candidates for the development of light-harvesting systems for potential applications in photovoltaics (Endo *et al.*, 2006, 2007; Ma *et al.*, 2008; Miller *et al.*, 2007).

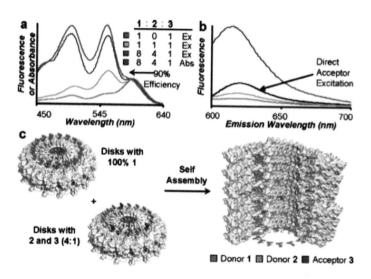

Figure 4.21 Three-chromophore systems for broad-spectrum light harvesting. TMV coat protein monomers labeled with 1 (Oregon Green), 2 (rhodamine), or 3 (Alexa Fluor 594) were combined in the ratios indicated in A. (A) Fluorescence excitation spectra (Ex), normalized at the acceptor excitation at 597 nm, indicated light harvesting over a wide range of wavelengths. The absorbance spectrum (Abs) for the 8 : 4 : 1 system is shown in red. (B) The antenna effect for each spectrum (λ_{ex} = 495 nm) is shown relative to the sample's acceptor emission by direct excitation (λ_{ex} = 588 nm). (C) Spatial distribution of chromophores for the 8 : 4 : 1 system. Reproduced with permission from Miller, R. A., Presley, A. D., and Francis, M. B. (2007) Self-assembling light-harvesting systems from synthetically modified tobacco mosaic virus coat proteins, *J. Am. Chem. Soc.*, **129**(11), 3104–3109.

4.5 PRACTICAL SELECTION GUIDE FOR VNPs

Many different VNPs have been utilized, and various bioconjugation protocols have been applied that allowed the attachment of a large variety of functional groups. Some highlights were discussed in this chapter. The following table gives a comprehensive overview of the different functionalities that have been attached to VNPs as of December 2009; this list is continuously expanding. Table 4.1 should be considered a snapshot of the various functionalities that have been conjugated to VNPs.

Table 4.1 Overview over the functionalities attached to VNPs

VNP	Target residue (E, exterior; I, interior; GE, genetically engineered)	Functional molecule(s) attached	Coupling reagent(s), type of chemistry	(Potential) Application(s)	Reference(s)
CCMV	Lys (E) Asp/Glu (E) Cys (E; GE)	Fluorescent dyes	NHS ester, homobifunctional linker, EDC/NHS, maleimide	Proof-of-concept	Gillitzer *et al.* (2002)
CCMV	Lys (E)	Biotin and digoxigenin	NHS ester	Breaking the symmetry and mixed assembly of multifunctional VNPs	Gillitzer *et al.* (2006)
CCMV	Cys (E; GE)	Iodoacetamide	Solid-state approach	Breaking the symmetry and controlled immobilization of VNPs	Klem *et al.* (2003)
CCMV	Lys (E)	Antibodies and fluorescent dyes or chelated gadolinium complexes	NHS ester and non-covalent biotin–streptavidin linkage	Biomedical targeting and imaging	Suci *et al.* (2007a,b)
CCMV	Lys (E) Cys (E; GE)	Antibodies and ruthenium complexes	NHS ester and non-covalent biotin–streptavidin linkage, maleimide	Targeted drug delivery	Suci *et al.* (2007a,b)
CCMV	Lys (E)	Biotin	NHS ester	Fabrication of multilayer arrays	Suci *et al.* (2006)
CPMV	Lys (E) Cys (E; GE) Cys (I)	Fluorescent dyes, biotin, nanogold, and small mercury compounds	Maleimide, haloacetyl derivatives, NHS ester, isothiocyanate, NHS-azide or -alkyne followed by click chemistry	Proof-of-concept	Wang *et al.* (2002a,b,c, 2003a), Chatterji *et al.* (2004a)
CPMV	Tyr (E)	Fluorescent dye and intraparticle cross-linking	One-electron and photochemical oxidation	Proof-of-concept	Meunier *et al.* (2004)

CPMV	Cys (I) Cys (E; GE) Lys (E)	Stilbene, fluorescent dyes, PEG	Bromoacetamide, NHS ester	Proof-of-concept, structural studies	Wang *et al.* (2003b)
CPMV	Lys (E)	Protected thiols	NHS ester	Alternative to genetically engineered mutants	Steinmetz *et al.* (2007)
CPMV	Lys (E)	Fluorescent dyes, NIR probes, PEG	NHS ester	Development of CPMV for intravital vascular imaging	Lewis *et al.* (2006), Wu *et al.* (2005)
CPMV	Lys (E)	Chelated gadolinium complexes	NHS-azide followed by click reactions	Contrast agents for MRI	Prasuhn *et al.* (2007)
CPMV	Lys (E)	QDs	EDC/NHS	Imaging and sensing devices	Portney *et al.* (2005)
CPMV	Hexa-His tag (E; GE)	QDs	Non-covalent Hexa-His–Ni-NTA interaction	Biosensors	Medintz *et al.* (2005)
CPMV	Cys (E; GE) Lys (E)	Iron oxide nanoparticles	Solid-state approach using maleimide and EDC/NHS coupling	Hybrids for potential MRI imaging	Martinez-Morales *et al.* (2008)
CPMV	Lys (E)	PEG	NHS ester, NHS-azide or -alkyne followed by click chemistry	PEGylation for biomedical applications	Sen Gupta *et al.* (2005a), Raja *et al.* (2003b), Steinmetz & Manchester (2009)
CPMV	Lys (E)	Buckyball (C_{60})	EDC/NHS	Photo-activated tumor therapy	Steinmetz *et al.* (2009b)
CPMV	Lys (E) Cys (E; GE)	Proteins - T4 lysozyme - LRR internalin B - Intron 8 of herstatin - Transferrin	Homo- and heterofunctional linkers (NHS ester and maleimide); NHS-azide followed by click chemistry	Biomedical targeting approaches	Chatterji *et al.* (2004b), Sen Gupta et al. (2005a)
CPMV	Lys (E)	Peptides: RGD and protective antigen (PA) of anthrax toxin	NHS-azide or -alkyne followed by click chemistry	Development of CPMV for biomedical approaches	Sen Gupta *et al.* (2005a)
CPMV	Lys (E)	Folic acid and PEG	NHS alkyne followed by click chemistry	Biomedical targeting approaches	Destito *et al.* (2007)

CPMV	Lys (E)	Carbohydrates	Isothiocyanate derivatives; NHS-alkyne followed by click reaction	Biomedical targeting approaches	Kaltgrad *et al.* (2008), Raja *et al.* (2003a), Sen Gupta *et al.* (2005a,b)
CPMV	Hexa-His (E; GE)	Nanogold	Non-covalent hexa-His–Ni-NTA interactions	Biosensing and nanoelctronic devices	Chatterji *et al.* (2005)
CPMV	Cys (E; GE)	Nanogold	Gold-thiol or maleimide	Biosensing and nanoelectronic devices	Soto *et al.* (2004), Blum *et al.* (2004, 2005a)
CPMV	Lys (E) Asp, Glu (E)	Redox-active compounds: ferrocene and viologen	EDC/NHS	Redox-active materials for catalysis and sensing devices	Steinmetz *et al.* (2006b,c)
CPMV	Cys (E; GE)	Avidin and fluorescent dyes	Maleimide	Probes and tracers for biosensing devices and microarray technology	Soto *et al.* (2006)
(CPMV	Cys (E; GE) Lys (exterior)	Antibodies and fluorescent dyes	Maleimide, heterobivalent linkers (NHS ester and maleimide)	Tracers in immunoassays and biosensors	Sapsford *et al.* (2006)
CPMV	Lys (E)	Biotin	Non-covalent biotin–streptavidin interactions	Fabrication of multilayered arrays	Steinmetz *et al.* (2006a, 2008b)
CPMV	Lys (E)	Oligonucleotides	Homobivalent linkers (NHS ester)	Self-assembling devices	Strable *et al.* (2004)
CPMV	Lys (E) Cys (E; GE)	Avidin, antibodies, and fluorescent dyes	Heterobivalent linkers (NHS and maleimide), NHS ester	VNPs as reporter tags in microtubule gliding assay	Martin *et al.* (2006)
HCRSV	Lys (E)	Folic acid	EDC/NHS	Targeted drug delivery devices	Ren *et al.* (2007)
TYMV	Lys (E) Asp (E)	Fluorescent dyes	EDC/NHS	Proof-of-concept	Barnhill *et al.* (2007b)
TYMV	Lys (E) Asp (E)	Terbium complexes, biotin	EDC/NHS	Time-resolved fluoro-immunoassay	Barnhill *et al.* (2007a)
FHV	Lys (E; GE)	SWCNTs	EDC/NHS	Imaging and sensing devices	Portney *et al.* (2005)
FHV	Lys (E; GE) Cys (E; GE)	Fluorescent dyes and targeting peptides	Isothiocyanate, NHS ester, and maleimide	Biomedical imaging and targeting	Destito *et al.* (2009)

P22	Hexa-His tag (E; GE)	Nanogold	Non-covalent Hexa-his–Ni-NTA interaction	Proof-of-concept	Kang *et al.* (2008)
MS2	Tyr (I)	Small chemicals	Diazonium coupling	Proof-of-concept	Hooker *et al.* (2004)
MS2	Cys (E; GE)	Fluorescent dye	Maleimide	Proof-of-concept	Peabody (2003)
MS2	Lys (E)	Fluorescent dyes, stearic acid	NHS ester, pentafluophenyl ester	Stabilization and solubilization of MS2 in organic solvents	Johnson *et al.* (2007)
MS2	Lys (E)	Antibodies, transferrin (therapeutic compounds were encapsulated)	NHS ester- and maleimide-reactive linkers	Targeted drug-delivery	Brown *et al.* (2002)
MS2	Lys (E)	TAT (cell penetrating peptide), (antisense therapeutic RNA was encapsidated)	NHS ester- and maleimide-reactive linkers	Targeted RNA delivery for gene silencing	Wei *et al.* (2009)
MS2	Tyr (I) Lys (E)	Chelated gadolinium complexes	Isothiocyanate or diazonium coupling, NHS-aldehyde followed by oxime condensation	Contrast agents for MRI	Hooker *et al.* (2007), Datta *et al.* (2008), Anderson *et al.* (2006)
MS2	Tyr (I)	[^{18}F]fluoro-benzaldehyde and fluorescent dyes	Diazonium coupling, followed by oxime condensation	Imaging platforms for positron emission tomography	Hooker *et al.* (2008)
MS2	Tyr (I) Lys (E)	Fluoresecent dyes, biotin, PEG-biotin, PEG-dye conjugates	Diazonium NHS-aldehyde followed by oxime condensation	Development of the MS2 platform for potential biomedical applications	Kovacs *et al.* (2007)
Qβ	Lys (E)	Fluorescent dye, bovine serum albumin (protein)	NHS-azide followed by click reaction	Optimized click chemistry protocol	Hong *et al.* (2009)
Qβ	Azide (E; GE) Alkyne (E; GE)	Fluorescent dyes, chelated gadolinium complexes, biotin	Click chemistry	Imaging modalities for biomedical applications	Strable *et al.* (2008)

Qβ	Lys (E)	Chelated gadolinium complexes	NHS-azide followed by click reactions	Contrast agents for MRI	Prasuhn *et al.* (2007, 2008)
Qβ	Lys (E)	Carbohydrates	NHS-alkyne followed by click reaction	Biomedical targeting approaches	Kaltgrad *et al.* (2008)
Qβ	Lys (E)	Polycationic motifs (poly Arg peptides)	NHS-azide and -alkyne followed by click reaction	Heparin antagonist for biomedical applications	Udit *et al.* (2009)
Qβ	Lys (E)	Nicotine	NHS ester	Nicotine vaccine for smoking cessation	Maurer *et al.* (2005)
Ad[1]	Lys (E)	PEG	NHS ester	Development of improved Ads for gene delivery with reduced immunogenicity and clearance	For reviews see: Singh and Kostarelos (2009), Barnett *et al.* (2002), Campos and Barry (2007), Mizuguchi and Hayakawa (2004)
A	Lys (E)	PEG-RGD	NHS ester	Retargeting Ads for tissue-specific gene delivery	Eto *et al.* (2005, 2008)
Ad	Lys (E)	PEG-folic acid and fluorescent dyes	EDC/NHS	Retargeting Ads for tissue-specific gene delivery	Oh *et al.* (2006)
Ad	Lys (E)	PEG-FGF	Heterobivalent linker (maleimide and NHS ester)	Retargeting Ads for tissue-specific gene delivery	Menezes *et al.* (2006)
Ad	Lys (E)	Carbohydrates	2-Imino-2-methoxyethyl-1-thioglycosides	Retargeting Ads for tissue-specific gene delivery	Pearce *et al.* (2005)
Ad	Cys (E; GE)	Nanogold, PEG, transferrin	Maleimide, NHS ester, and heterobivalent linker	Retargeting of Ad and delivery of gold nanoparticles for biomedical applications	Kreppel *et al.* (2005)

Ad	Lys (E)	Phthalocyanines	EDC/NHS	Using Ad as a photosensitizer for photodynamic therapy	Allen *et al.* (1999)
Ad	Lys (E)	Nanogold and anti-CEA fragment (MFE)	NHS ester Non-covalent attachment of sCAR-MFE	Hyperthermia treatment for tumor therapy	Everts *et al.* (2006)
Ad	Hexa-His tag (E; GE)	Nanogold and anti-CEA fragment (MFE)	Non-covalent hexa-His–Ni-NTA interactions and non-covalent attachment of sCAR-MFE	Hyperthermia treatment for tumor therapy	Saini *et al.* (2008)
CPV	Lys (E)	Fluorescent dyes	NHS ester	Development of CPV for biomedical targeting applications	Singh *et al.* (2006)
PVX	Lys (E)	Fluorescent dyes and PEG	NHS ester; NHS-alkyne followed by click chemistry	Development of PVX for biomedical applications	Steinmetz *et al.* (2009c)
TMV	Lys (E; GE)	Biotin and fluorescent dyes	NHS ester	Proof-of-concept	Demir and Stockwell (2002)
TMV	Tyr (E) Glu (I)	Biotin, PEG, small chemicals, and fluorescent dyes	Diazonium coupling, oxime condensation, EDC/NHS	Proof-of-concept	Schlick *et al.* (2005)
TMV	Tyr (E)	PEG and RGD	Diazonium (aniline–alkyne) followed by click chemistry	Development of TMV for biomedical applications	Bruckman *et al.* (2008)
TMV	Cys (E; GE)	Fluorescent dyes	Maleimide	Immobilization and fabrication of microarrays	Yi *et al.* (2005, 2007)
TMV	Cys (I; GE)	Fluorescent dyes, porphyrins or pyrene	Maleimide	Light-harvesting systems	Ma *et al.* (2008), Endo *et al.* (2006, 2007), Miller *et al.* (2007)
M13	N-term of pVIII	NIR fluorophores	NHS ester	VNP-based probes for biomedical imaging	Hilderbrand *et al.* (2005, 2008)
M13	Amines (Lys and amino-terminus; not specified)	Iron oxide nanoparticles	Epichlorhydrin	Magnetophage for MRI	Segers *et al.* (2007)

M13	Glu/Asp (E; pVIII)	Cytotoxic drugs (doxorubicin and hygromycin) and antibodies (anti-ERGR and anti-ErbB2)	EDC as covalent strategy Antibodies were attached via a non-covalent strategy using a genetically introduced IgG-binding domain	M13 as a tool for targeted drug delivery to cancer cells	Bar *et al.* (2008)
M13	Amines (Lys and amino-terminus; not specified)	Chloramphenicol and targeting peptide (genetically introduced)	NHS ester	Targeted drug delivery to bacteria	Yacoby *et al.* (2006)
M13	Lys (E; on pVIII)	Polyacrylamide polymer	Traut's reagent to introduce thiol functional group followed by maleimide coupling	Orderd nanoscale arrays via self-assembly	Willis *et al.* (2008)
M13	Hexa-His Tag (GE; pIX) Selenocysteine (GE; pIII)	Biotin	Iodoacetyl-PEG-linker	Tethering and testing biopolymer elasticity using a AFM approach	Khalil *et al.* (2007a)
SIRV2	Lys (E, at the capsid end structures) Asp/Glu (E) carbohydrates (E)	Fluorescent dyes and biotin	NHS ester, EDC/NHS, mild oxidation and hydrazide coupling	Proof-of-concept	Steinmetz *et al.* (2008a)

Applications and fields are color coded: pale green, chemistry/proof-of-concept study; light blue, biomedicine; lavender, materials.

[1]Adenoviruses as well as AAV have been studied and utilized as gene delivery vectors for many years. Genetic as well as chemical modification strategies have been extensively studied and explored by many research groups, and the literature is tremendous. It is not possible to capture in this table all the efforts made toward bioconjugation on adenoviruses, the reader is referred to the references and reviews given throughout the book.

References

Allen, C. M., Sharman, W. M., La Madeleine, C., Weber, J. M., Langlois, R., Ouellet, R., and van Lier, J. E. (1999) Photodynamic therapy: tumor targeting with adenoviral proteins, *Photochem. Photobiol.*, **70**(4), 512–523.

Anderson, E. A., Isaacman, S., Peabody, D. S., Wang, E. Y., Canary, J. W., and Kirshenbaum, K. (2006) Viral nanoparticles donning a paramagnetic coat: conjugation of MRI contrast agents to the MS2 capsid, *Nano Lett.*, **6**(6), 1160–1164.

Aslam, M., and Dent, A. (1999) *Bioconjugation*, Macmillan Reference Ltd., London.

Baker, A. H. (2003) Targeting AAV vectors, *Mol. Ther.*, **7**(4), 433–434.

Bar, H., Yacoby, I., and Benhar, I. (2008) Killing cancer cells by targeted drug-carrying phage nanomedicines, *BMC Biotechnol.*, **8**, 37.

Barnett, B. G., Crews, C. J., and Douglas, J. T. (2002) Targeted adenoviral vectors, *Biochim. Biophys. Acta*, **1575**(1–3), 1–14.

Barnhill, H. N., Claudel-Gillet, S., Ziessel, R., Charbonniere, L. J., and Wang, Q. (2007a) Prototype protein assembly as scaffold for time-resolved fluoroimmuno assays, *J. Am Chem Soc* **129**(25), 7799–7806.

Barnhill, H. N., Reuther, R., Ferguson, P. L., Dreher, T., and Wang, Q. (2007b) Turnip yellow mosaic virus as a chemoaddressable bionanoparticle, *Bioconjug. Chem.*, **18**(3), 852–859.

Blum, A. S., Soto, C. M., Wilson, C. D., Brower, T. L., Pollack, S. K., Schull, T. L., Chatterji, A., Lin, T., Johnson, J. E., Amsinck, C., Franzon, P., Shashidhar, R., and Ratna, B. R. (2005a) An engineered virus as a scaffold for three-dimensional self-assembly on the nanoscale, *Small*, **1**(7), 702–706.

Blum, A. S., Soto, C. M., Wilson, C. D., Cole, J. D., Kim, M., Gnade, B., Chatterji, A., Ochoa, W. F., Lin, T., Johnson, J. E., and Ratna, B. R. (2004) Cowpea mosaic virus as a scaffold for 3-D patterning of gold nanoparticles, *Nano Lett.*, **4**, 867–870.

Brown, W. L., Mastico, R. A., Wu, M., Heal, K. G., Adams, C. J., Murray, J. B., Simpson, J. C., Lord, J. M., Taylor-Robinson, A. W., and Stockley, P. G. (2002) RNA bacteriophage capsid-mediated drug delivery and epitope presentation, *Intervirology*, **45**(4–6), 371–380.

Bruckman, M. A., Kaur, G., Lee, L. A., Xie, F., Sepulveda, J., Breitenkamp, R., Zhang, X., Joralemon, M., Russell, T. P., Emrick, T., and Wang, Q. (2008) Surface modification of tobacco mosaic virus with "click" chemistry, *ChemBioChem*, **9**(4), 519–523.

Campos, S. K., and Barry, M. A. (2007) Current advances and future challenges in Adenoviral vector biology and targeting, *Curr. Gene Ther.*, **7**(3), 189–204.

Chatterji, A., Ochoa, W., Paine, M., Ratna, B. R., Johnson, J. E., and Lin, T. (2004a) New addresses on an addressable virus nanoblock: uniquely reactive Lys residues on cowpea mosaic virus, *Chem. Biol.*, **11**(6), 855–863.

Chatterji, A., Ochoa, W., Shamieh, L., Salakian, S. P., Wong, S. M., Clinton, G., Ghosh, P., Lin, T., and Johnson, J. E. (2004b) Chemical conjugation of heterologous proteins on the surface of Cowpea mosaic virus, *Bioconjug. Chem.*, **15**(4), 807–813.

Chatterji, A., Ochoa, W. F., Ueno, T., Lin, T., and Johnson, J. E. (2005) A virus-based nanoblock with tunable electrostatic properties, *Nano Lett.*, **5**(4), 597–602.

Datta, A., Hooker, J. M., Botta, M., Francis, M. B., Aime, S., and Raymond, K. N. (2008) High relaxivity gadolinium hydroxypyridonate-viral capsid conjugates: nanosized MRI contrast agents, *J. Am. Chem. Soc.*, **130**(8), 2546–2552.

Demir, M., and Stockwell, M. H. B. (2002) A chemoselective biomolecular template for assembling diverse nanotubular materials, *Nanotechnology*, **13**, 541–544.

Destito, G., Schneemann, A., and Manchester, M. (2009) Biomedical nanotechnology using virus-based nanoparticles, *Curr. Top. Microbiol. Immunol.*, **327**, 95–122.

Destito, G., Yeh, R., Rae, C. S., Finn, M. G., and Manchester, M. (2007) Folic acid-mediated targeting of cowpea mosaic virus particles to tumor cells, *Chem. Biol.*, **14**(10), 1152–1162.

Dirksen, A., and Dawson, P. E. (2008) Rapid oxime and hydrazone ligations with aromatic aldehydes for biomolecular labeling, *Bioconjug. Chem.*, **19**(12), 2543–2548.

Endo, M., Fujitsuka, M., and Majima, T. (2007) Porphyrin light-harvesting arrays constructed in the recombinant tobacco mosaic virus scaffold, *Chemistry*, **13**(31), 8660–8666.

Endo, M., Wang, H., Fujitsuka, M., and Majima, T. (2006) Pyrene-stacked nanostructures constructed in the recombinant tobacco mosaic virus rod scaffold, *Chemistry*, **12**(14), 3735–3740.

Eto, Y., Gao, J. Q., Sekiguchi, F., Kurachi, S., Katayama, K., Maeda, M., Kawasaki, K., Mizuguchi, H., Hayakawa, T., Tsutsumi, Y., Mayumi, T., and Nakagawa, S. (2005) PEGylated adenovirus vectors containing RGD peptides on the tip of PEG show high transduction efficiency and antibody evasion ability, *J. Gene Med.*, **7**(5), 604–612.

Eto, Y., Yoshioka, Y., Mukai, Y., Okada, N., and Nakagawa, S. (2008) Development of PEGylated adenovirus vector with targeting ligand, *Int. J. Pharm.*, **354**(1–2), 3–8.

Everts, M., Saini, V., Leddon, J. L., Kok, R. J., Stoff-Khalili, M., Preuss, M. A., Millican, C. L., Perkins, G., Brown, J. M., Bagaria, H., Nikles, D. E., Johnson, D. T., Zharov, V. P., and Curiel, D. T. (2006) Covalently linked Au nanoparticles to a viral vector: potential for combined photothermal and gene cancer therapy, *Nano Lett.*, **6**(4), 587–591.

Floyd, N., Vijayakrishnan, B., Koeppe, J. R., and Davis, B. G. (2009) Thiyl glycosylation of olefinic proteins: S-linked glycoconjugate synthesis, *Angew. Chem. Int. Ed. Engl.*, **48**(42), 7798–7802.

Gillitzer, E., Suci, P., Young, M., and Douglas, T. (2006) Controlled ligand display on a symmetrical protein-cage architecture through mixed assembly, *Small*, **2**(8–9), 962–966.

Gillitzer, E., Willits, D., Young, M., and Douglas, T. (2002) Chemical modification of a viral cage for multivalent presentation, *Chem. Comm.*, (20), 2390–2391.

Golmohammadi, R., Fridborg, K., Bundule, M., Valegard, K., and Liljas, L. (1996) The crystal structure of bacteriophage Q beta at 3.5 A resolution, *Structure*, **4**(5), 543–554.

Hermanson, G. T. (1996) *Bioconjugate Techniques*, Academic Press, Elsevier, London.

Hilderbrand, S. A., Kelly, K. A., Niedre, M., and Weissleder, R. (2008) Near infrared fluorescence-based bacteriophage particles for ratiometric pH imaging, *Bioconjug. Chem.*, **19**(8), 1635–1639.

Hilderbrand, S. A., Kelly, K. A., Weissleder, R., and Tung, C. H. (2005) Monofunctional near-infrared fluorochromes for imaging applications, *Bioconjug. Chem.*, **16**(5), 1275–1281.

Hong, V., Presolski, S. I., Ma, C., and Finn, M. G. (2009) Analysis and optimization of copper-catalyzed azide-alkyne cycloaddition for bioconjugation, *Angew. Chem. Int. Ed. Engl.*, **48**(52), 9879–9883.

Hooker, J. M., Datta, A., Botta, M., Raymond, K. N., and Francis, M. B. (2007) Magnetic resonance contrast agents from viral capsid shells: a comparison of exterior and interior cargo strategies, *Nano Lett.*, **7**(8), 2207–2210.

Hooker, J. M., Kovacs, E. W., and Francis, M. B. (2004) Interior surface modification of bacteriophage MS2, *J. Am. Chem. Soc.*, **126**(12), 3718–3719.

Hooker, J. M., O'Neil, J. P., Romanini, D. W., Taylor, S. E., and Francis, M. B. (2008) Genome-free viral capsids as carriers for positron emission tomography radiolabels, *Mol. Imaging Biol.*, **10**(4), 182–191.

Johnson, H. R., Hooker, J. M., Francis, M. B., and Clark, D. S. (2007) Solubilization and stabilization of bacteriophage MS2 in organic solvents, *Biotechnol. Bioeng.*, **97**(2), 224–234.

Kaltgrad, E., O'Reilly, M. K., Liao, L., Han, S., Paulson, J. C., and Finn, M. G. (2008) On-virus construction of polyvalent glycan ligands for cell-surface receptors, *J. Am. Chem. Soc.*, **130**(14), 4578–4579.

Kang, S., Lander, G. C., Johnson, J. E., and Prevelige, P. E. (2008) Development of bacteriophage p22 as a platform for molecular display: genetic and chemical modifications of the procapsid exterior surface, *ChemBioChem*, **9**(4), 514–518.

Khalil, A. S., Ferrer, J. M., Brau, R. R., Kottmann, S. T., Noren, C. J., Lang, M. J., and Belcher, A. M. (2007a) Single M13 bacteriophage tethering and stretching, *Proc. Natl Acad. Sci. USA*, **104**(12), 4892–4897.

Klem, M. T., Willits, D., Young, M., and Douglas, T. (2003) 2-D array formation of genetically engineered viral cages on Au surfaces and imaging by atomic force microscopy, *J. Am. Chem. Soc.*, **125**(36), 10806–10807.

Kolb, H. C., Finn, M. G., and Sharpless, K. B. (2001) Click chemistry: diverse chemical function from a few good reactions, *Angew. Chem. Int. Ed. Engl.*, **40**(11), 2004–2021.

Kovacs, E. W., Hooker, J. M., Romanini, D. W., Holder, P. G., Berry, K. E., and Francis, M. B. (2007) Dual-surface-modified bacteriophage MS2 as an ideal scaffold for a viral capsid-based drug delivery system, *Bioconjug. Chem.*, **18**(4), 1140–1147.

Kreppel, F., Gackowski, J., Schmidt, E., and Kochanek, S. (2005) Combined genetic and chemical capsid modifications enable flexible and efficient de- and retargeting of adenovirus vectors, *Mol. Ther.*, **12**(1), 107–117.

Kwon, I., and Schaffer, D. V. (2008) Designer gene delivery vectors: molecular engineering and evolution of adeno-associated viral vectors for enhanced gene transfer. *Pharm. Res.*, **25**(3), 489–499.

Lewis, J. D., Destito, G., Zijlstra, A., Gonzalez, M. J., Quigley, J. P., Manchester, M., and Stuhlmann, H. (2006) Viral nanoparticles as tools for intravital vascular imaging, *Nat. Med.*, **12**(3), 354–360.

Lin, T., Chen, Z., Usha, R., Stauffacher, C. V., Dai, J. B., Schmidt, T., and Johnson, J. E. (1999) The refined crystal structure of cowpea mosaic virus at 2.8 A resolution, *Virology*, **265**(1), 20–34.

Ma, Y. Z., Miller, R. A., Fleming, G. R., and Francis, M. B. (2008) Energy transfer dynamics in light-harvesting assemblies templated by the tobacco mosaic virus coat protein, *J. Phys. Chem. B*, **112**(22), 6887–6892.

Martin, B. D., Soto, C. M., Blum, A. S., Sapsford, K. E., Whitley, J. L., Johnson, J. E., Chatterji, A., and Ratna, B. R. (2006) An engineered virus as a bright fluorescent tag and scaffold for cargo proteins – capture and transport by gliding microtubules, *J. Nanosci. Nanotechnol.*, **6**, 2451–2460.

Martinez-Morales, A. A., Portney, N. G., Zhang, Y., Destito, G., Budak, G., Ozbay, E., Manchester, M., Ozkan, C. S., and Ozkan, M. (2008) Martinez-Morales, Nathaniel G. Portney, Yu Zhang, Giuseppe Destito, Gurer Budak, Ekmel Ozbay, Marianne Manchester, Cengiz S. Ozkan, and Mihrimah Ozkan, *Adv. Mater.*, **20**, 1–5.

Maurer, P., Jennings, G. T., Willers, J., Rohner, F., Lindman, Y., Roubicek, K., Renner, W. A., Muller, P., and Bachmann, M. F. (2005) A therapeutic vaccine for nicotine dependence: preclinical efficacy, and phase I safety and immunogenicity, *Eur. J. Immunol.*, **35**(7), 2031–2040.

Medintz, I. L., Sapsford, K. E., Konnert, J. H., Chatterji, A., Lin, T., Johnson, J. E., and Mattoussi, H. (2005) Decoration of discretely immobilized cowpea mosaic virus with luminescent quantum dots, *Langmuir*, **21**(12), 5501–5510.

Menezes, K. M., Mok, H. S., and Barry, M. A. (2006) Increased transduction of skeletal muscle cells by fibroblast growth factor-modified adenoviral vectors, *Hum. Gene Ther.*, **17**(3), 314–320.

Meunier, S., Strable, E., and Finn, M. G. (2004) Crosslinking of and coupling to viral capsid proteins by tyrosine oxidation, *Chem. Biol.*, **11**(3), 319–326.

Miller, R. A., Presley, A. D., and Francis, M. B. (2007) Self-assembling light-harvesting systems from synthetically modified tobacco mosaic virus coat proteins, *J. Am. Chem. Soc.*, **129**(11), 3104–3109.

Mizuguchi, H., and Hayakawa, T. (2004) Targeted adenovirus vectors, *Hum. Gene Ther.*, **15**(11), 1034–1044.

Nam, K. T., Peelle, B. R., Lee, S. W., and Belcher, A. M. (2004) Genetically driven assembly of nanorings based on the M13 virus, *Nano Lett.*, **4**(1), 23–27.

Nelson, N., and Ben-Shem, A. (2004) The complex architecture of oxygenic photosynthesis, *Nat. Rev. Mol. Cell. Biol.*, **5**(12), 971–982.

Oh, I. K., Mok, H., and Park, T. G. (2006) Folate immobilized and PEGylated adenovirus for retargeting to tumor cells, *Bioconjug. Chem.*, **17**(3), 721–727.

Peabody, D. S. (2003) A viral platform for chemical modification and multivalent display, *J. Nanobiotechnol.*, **1**(1), 5.

Pearce, O. M., Fisher, K. D., Humphries, J., Seymour, L. W., Smith, A., and Davis, B. G. (2005) Glycoviruses: chemical glycosylation retargets adenoviral gene transfer, *Angew. Chem. Int. Ed. Engl.*, **44**(7), 1057–1061.

Portney, N. G., Singh, K., Chaudhary, S., Destito, G., Schneemann, A., Manchester, M., and Ozkan, M. (2005) Organic and inorganic nanoparticle hybrids, *Langmuir*, **21**(6), 2098–2103.

Prasuhn, D. E., Jr., Singh, P., Strable, E., Brown, S., Manchester, M., and Finn, M. G. (2008) Plasma clearance of bacteriophage Qbeta particles as a function of surface charge, *J. Am. Chem. Soc.*, **130**(4), 1328–1334.

Prasuhn, D. E., Jr., Yeh, R. M., Obenaus, A., Manchester, M., and Finn, M. G. (2007) Viral MRI contrast agents: coordination of Gd by native virions and attachment of Gd complexes by azide–alkyne cycloaddition, *Chem. Commun.*, (12), 1269–1271.

Raja, K. S., Wang, Q., and Finn, M. G. (2003a) Icosahedral virus particles as polyvalent carbohydrate display platforms, *ChemBioChem*, **4**(12), 1348–1351.

Raja, K. S., Wang, Q., Gonzalez, M. J., Manchester, M., Johnson, J. E., and Finn, M. G. (2003b) Hybrid virus-polymer materials. 1. Synthesis and properties of PEG-decorated cowpea mosaic virus, *Biomacromolecules*, **3**, 472–476.

Ren, Y., Wong, S. M., and Lim, L. Y. (2007) Folic acid-conjugated protein cages of a plant virus: a novel delivery platform for doxorubicin, *Bioconjug. Chem.*, **18**(3), 836–843.

Saini, V., Martyshkin, D. V., Mirov, S. B., Perez, A., Perkins, G., Ellisman, M. H., Towner, V. D., Wu, H., Pereboeva, L., Borovjagin, A., Curiel, D. T., and Everts, M. (2008) An adenoviral platform for selective self-assembly and targeted delivery of nanoparticles, *Small*, **4**(2), 262–269.

Sapsford, K. E., Soto, C. M., Blum, A. S., Chatterji, A., Lin, T., Johnson, J. E., Ligler, F. S., and Ratna, B. R. (2006) A cowpea mosaic virus nanoscaffold for multiplexed antibody conjugation: application as an immunoassay tracer, *Biosens. Bioelectron.*, **21**(8), 1668–1673.

Scheck, R. A., Dedeo, M. T., Iavarone, A. T., and Francis, M. B. (2008) Optimization of a biomimetic transamination reaction, *J. Am. Chem. Soc.*, **130**(35), 11762–11770.

Schlick, T. L., Ding, Z., Kovacs, E. W., and Francis, M. B. (2005) Dual-surface modification of the tobacco mosaic virus, *J. Am. Chem. Soc.*, **127**, 3718–3723.

Segers, J., Laumonier, C., Burtea, C., Laurent, S., Elst, L. V., and Muller, R. N. (2007) From phage display to magnetophage, a new tool for magnetic resonance molecular imaging, *Bioconjug. Chem.*, **18**(4), 1251–1258.

Sen Gupta, S., Kuzelka, J., Singh, P., Lewis, W. G., Manchester, M., and Finn, M. G. (2005a) Accelerated bioorthogonal conjugation: a practical method for the ligation of diverse functional molecules to a polyvalent virus scaffold, *Bioconjug. Chem.*, **16**(6), 1572–1579.

Sen Gupta, S., Raja ,K. S., Kaltgrad, E., Strable, E., and Finn, M. G. (2005b) Virus-glycopolymer conjugates by copper(I) catalysis of atom transfer radical polymerization and azide–alkyne cycloaddition, *Chem. Commun. (Camb)*, (34), 4315–4317.

Singh, P., Destito, G., Schneemann, A., and Manchester, M. (2006) Canine parvovirus-like particles, a novel nanomaterial for tumor targeting, *J. Nanobiotechnol.*, **4**, 2.

Singh, R., and Kostarelos, K. (2009) Designer adenoviruses for nanomedicine and nanodiagnostics, *Trends Biotechnol.*, **27**(4), 220–229.

Soto, C. M., Blum, A. S., Vora, G. J., Lebedev, N., Meador, C. E., Won, A. P., Chatterji, A., Johnson, J. E., and Ratna, B. R. (2006) Fluorescent signal amplification of carbocyanine dyes using engineered viral nanoparticles, *J. Am. Chem. Soc.*, **128**(15), 5184–5189.

Soto, C. M., Blum, A. S., Wilson, C. D., Lazorcik, J., Kim, M., Gnade, B., and Ratna, B. R. (2004) Separation and recovery of intact gold-virus complex by agarose electrophoresis and electroelution: application to the purification of cowpea mosaic virus and colloidal gold complex, *Electrophoresis*, **25**(17), 2901–2906.

Steinmetz, N. F., Bize, A., Findlay, K. C., Lomonossoff, G. P., Manchester, M., Evans, D. J., and Prangishvili, D. (2008a) Site-specific and spatially controlled addressability of a new viral nanobuilding block: sulfolobus islandicus rod-shaped virus 2, *Adv. Funct. Mater.*, **18**, 3478–3486.

Steinmetz, N. F., Bock, E., Richter, R. P., Spatz, J. P., Lomonossoff, G. P., and Evans, D. J. (2008b) Assembly of multilayer arrays of viral nanoparticles via biospecific recognition: a quartz crystal microbalance with dissipation monitoring study, *Biomacromolecules*, 9(2), 456–462.

Steinmetz, N. F., Calder, G., Lomonossoff, G. P., and Evans, D. J. (2006a) Plant viral capsids as nanobuilding blocks: construction of arrays on solid supports, *Langmuir*, **22**(24), 10032–10037.

Steinmetz, N. F., Evans, D. J., and Lomonossoff, G. P. (2007) Chemical introduction of reactive thiols into a viral nanoscaffold: a method that avoids virus aggregation, *ChemBioChem*, **8**(10), 1131–1136.

Steinmetz, N. F., Hong, V., Spoerke, E. D., Lu, P., Breitenkamp, K., Finn, M. G., and Manchester, M. (2009a) Buckyballs meet viral nanoparticles – candidates for biomedicine, *J. Am. Chem. Soc.*, **31**(47), 17093–17095.

Steinmetz, N. F., Hong, V., Spoerke, E. D., Lu, P., Breitenkamp, K., Finn, M. G., and Manchester, M. (2009b) Buckyballs meet viral nanoparticles: candidates for biomedicine, *J. Am. Chem. Soc.*, **131**(47), 17093–17095.

Steinmetz, N. F., Lomonossoff, G. P., and Evans, D. J. (2006b) Cowpea mosaic virus for material fabrication: addressable carboxylate groups on a programmable nanoscaffold, *Langmuir*, **22**(8), 3488–3490.

Steinmetz, N. F., Lomonossoff, G. P., and Evans, D. J. (2006c) Decoration of cowpea mosaic virus with multiple, redox-active, organometallic complexes, *Small*, **2**(4), 530–533.

Steinmetz, N. F., and Manchester, M. (2009) PEGylated viral nanoparticles for biomedicine: the impact of PEG chain length on vnp cell interactions in vitro and ex vivo, *Biomacromolecules*, **10**(4), 784–792.

Steinmetz, N. F., Mertens, M. E., Taurog, R. E., Johnson, J. E., Commandeur, U., Fischer, R., and Manchester, M. (2009c) Potato virus X as a novel platform for potential biomedical applications, *Nano Lett.*, **10**(1), 305–312.

Strable, E., and Finn, M. G. (2009) Chemical modification of viruses and virus-like particles, *Curr. Top. Microbiol. Immunol.*, **327**, 1–21.

Strable, E., Johnson, J. E., and Finn, M. G. (2004) Natural nanochemical building blocks: icosahedral virus particles organized by attached oligonucleotides, *Nano Lett.*, **4**, 1385–1389.

Strable, E., Prasuhn, D. E., Jr., Udit, A. K., Brown, S., Link, A. J., Ngo, J. T., Lander, G., Quispe, J., Potter, C. S., Carragher, B., Tirrell, D. A., and Finn, M. G. (2008) Unnatural amino acid incorporation into virus-like particles, *Bioconjug. Chem.*, **19**(4), 866–875.

Suci, P. A., Berglund, D. L., Liepold, L., Brumfield, S., Pitts, B., Davison, W., Oltrogge, L., Hoyt, K. O., Codd, S., Stewart, P. S., Young, M., and Douglas, T. (2007a) High-density targeting of a viral multifunctional nanoplatform to a pathogenic, biofilm-forming bacterium, *Chem. Biol.*, **14**(4), 387–398.

Suci, P. A., Klem, M. T., Arce, F. T., Douglas, T., and Young, M. (2006) Assembly of multilayer films incorporating a viral protein cage architecture, *Langmuir*, **22**(21), 8891–8896.

Suci, P. A., Varpness, Z., Gillitzer, E., Douglas, T., and Young, M. (2007b) Targeting and photodynamic killing of a microbial pathogen using protein cage architectures functionalized with a photosensitizer, *Langmuir*, **23**(24), 12280–12286.

Udit, A. K., Brown, S., Baksh, M. M., and Finn, M. G. (2008) Immobilization of bacteriophage Qbeta on metal-derivatized surfaces via polyvalent display of hexahistidine tags, *J. Inorg. Biochem.*, **102**(12), 2142–2146.

Udit, A. K., Everett, C., Gale, A. J., Reiber Kyle, J., Ozkan, M., and Finn, M. G. (2009) Heparin antagonism by polyvalent display of cationic motifs on virus-like particles, *ChemBioChem*, **10**(3), 503–510.

Wang, Q., Chan, T. R., Hilgraf, R., Fokin, V. V., Sharpless, K. B., and Finn, M. G. (2003a) Bioconjugation by copper(I)-catalyzed azide–alkyne [3 + 2] cycloaddition, *J. Am. Chem. Soc.*, **125**(11), 3192–3193.

Wang, Q., Kaltgrad, E., Lin, T., Johnson, J. E., and Finn, M. G. (2002a) Natural supramolecular building blocks: wild-type cowpea mosaic virus, *Chem. Biol.*, **9**(7), 805–811.

Wang, Q., Lin, T., Johnson, J. E., and Finn, M. G. (2002b) Natural supramolecular building blocks: cysteine-added mutants of cowpea mosaic virus, *Chem. Biol.*, **9**(7), 813–819.

Wang, Q., Lin, T., Tang, L., Johnson, J. E., and Finn, M. G. (2002c) Icosahedral virus particles as addressable nanoscale building blocks, *Angew. Chem. Int. Ed.*, **41**(3), 459–462.

Wang, Q., Raja, K. S., Janda, K. D., Lin, T., and Finn, M. G. (2003b) Blue fluorescent antibodies as reporters of steric accessibility in virus conjugates, *Bioconjug. Chem.*, **14**(1), 38–43.

Wei, B., Wei, Y., Zhang, K., Wang, J., Xu, R., Zhan, S., Lin, G., Wang, W., Liu, M., Wang, L., Zhang, R., and Li, J. (2009) Development of an antisense RNA delivery system using conjugates of the MS2 bacteriophage capsids and HIV-1 TAT cell-penetrating peptide, *Biomed. Pharmacother.*, **63**(4), 313–318.

Willis, B., Eubanks, L. M., Wood, M. R., Janda, K. D., Dickerson, T. J., and Lerner, R. A. (2008) Biologically templated organic polymers with nanoscale order, *Proc. Natl Acad. Sci. USA*, **105**(5), 1416–1419.

Wu, C., Barnhill, H., Liang, X., Wang, Q., and Jiang, H. (2005) A nw probe using hybrid virus-dye nanoparticles for near-infrared fluorescence tomography, *Opt. Commun.*, **255**, 366–374.

Yacoby, I., Shamis, M., Bar, H., Shabat, D., and Benhar, I. (2006) Targeting antibacterial agents by using drug-carrying filamentous bacteriophages, *Antimicrob. Agents Chemother.*, **50**(6), 2087–2097.

Yi, H., Nisar, S., Lee, S. Y., Powers, M. A., Bentley, W. E., Payne, G. F., Ghodssi, R., Rubloff, G. W., Harris, M. T., and Culver, J. N. (2005) Patterned assembly of genetically modified viral nanotemplates via nucleic acid hybridization, *Nano Lett.*, **5**(10), 1931–1936.

Yi, H., Rubloff, G. W., and Culver, J. N. (2007) TMV microarrays: hybridization-based assembly of DNA-programmed viral nanotemplates, *Langmuir*, **23**(5), 2663–2667.

Chapter 5

ENCAPSULATING MATERIALS WITHIN VNPs

Viral nanoparticles (VNPs) contain three surfaces that can be exploited for functionalization: the exterior capsid surface, the interior, and the interface between adjacent coat proteins. The previous chapter dealt with the functionalization of either the exterior or interior surfaces using covalent bioconjugation strategies (see Chapter 4). This chapter will summarize different approaches that allow the encapsulation of materials within the interior cavity of the VNP.

In order to achieve encapsulation, a range of techniques may be applied:

1. *Infusion of small molecules*

   Although they often appear to be closed shells in structural representations, virus capsids contain pores on their surfaces of varying sizes. Not only do the capsids of VNPs contain pores, but they are also highly dynamic: structural transitions of the pentamers and hexamers occur as a result of thermal fluctuation (Bothner *et al.*, 1998; Gibbons & Klug, 2007; Witz & Brown, 2001). Small molecules can diffuse freely between the bulk medium and the capsid interior through these pores and openings. Once inside the interior cavity, the compound of interest can be covalently attached to addressable amino acids on the interior surface of the capsid. Covalent interior modification has been shown for *Cowpea mosaic virus* (CPMV), MS2, and *Tobacco mosaic virus* (TMV) (discussed in Section 4.3.7).

   Retention can also be achieved based on interactions with the encapsidated nucleic acids; these are typically non-covalent interactions. Small positively charged molecules or molecules with natural affinity to nucleic acids can be stably entrapped (see Section 5.1).

   An alternative strategy makes use of the so-called gating mechanisms; recall the pH- and metal ion-dependent structural transitions of, for

Viral Nanoparticles: Tools for Materials Science and Biomedicine
By Nicole F. Steinmetz and Marianne Manchester
Copyright © 2011 by Pan Stanford Publishing Pte. Ltd.
www.panstanford.com

example, *Brome mosaic virus* (BMV), *Cowpea chlorotic mottle virus* (CCMV), and *Red clover necrotic mottle virus* (RCNMV). Here, the material is infused into particles in the swollen form, that is, the "open" conformation. Dialysis against the appropriate buffers allows reversal of the swelling and the particles will return to the non-swollen "closed" conformation. The materials that diffused into the open capsids are thus entrapped (discussed in Section 5.1).

2. *Constrained materials synthesis*

   VNPs have been exploited as constrained reaction vessels. Here, precursor materials diffuse into the interior cavity of assembled VNPs, where nucleation and mineralization of the material occurs leading to the synthesis of inorganic nanocrystals. Materials synthesis within VNPs has been extensively studied using the icosahedron CCMV and the rod TMV; these techniques will be discussed in Chapter 6.

3. *Encapsulation during self-assembly*

   In vitro self-assembly mechanisms can be exploited to selectively entrap materials within a virus-like particle (VLP). Here, the particles are disassembled into coat protein monomers. The monomers are then mixed with the material of interest and re-assembled. If successful, the VLP assembles around the artificial cargo. Encapsulation of polymers, synthetic nanoparticles, and enzymes has been demonstrated using the self-assembly strategy and will be discussed in Section 5.2.

5.1 INFUSION OF MATERIALS INTO ASSEMBLED VNPs

One of the first major papers published in the field of viral nanotechnology dealt with the exploration of VNPs as nanocontainers to host artificial cargo and as constrained reaction vessels for materials synthesis (Douglas & Young, 1998). (The utilization of CCMV as a constrained reaction vessel will be discussed in Chapter 6.) The pH- and metal ion-dependent structural transitions of the CCMV capsid made it an obvious target to study host–guest encapsulation and the selective encapsulation of anionic organic polymers. In this case, polyanetholesulfonic acid (PASA) and polydextran sulfate (PDS) were packed into the CCMV cage (Douglas & Young, 1998). CCMV particles in the swollen, open conformation (pH > 6.5 and depletion of metal ions from the buffer) were exposed to the polymers. Under these conditions, the polymers have free access to the interior through the 60

2 nm-sized pores. Inside the viral capsid, the negatively charged polymers bind to the positively charged interior capsid surface via electrostatic interactions. Lowering of the pH (pH < 6.5) and addition of metal cations results in structural transition into the non-swollen, closed conformation; the polymers are trapped inside the VLP (Douglas & Young, 1998).

Small molecules can be infused and entrapped into VNPs by making use of the encapsidated cargo, the nucleic acid. For example, the lanthanides Gd^{3+} and Tb^{3+} have been successfully and stably entrapped within CPMV particles based on RNA interactions (Prasuhn *et al.*, 2007; Singh *et al.*, 2007). In brief, intact, wild-type CPMV particles were immersed in a solution containing Gd^{3+} or Tb^{3+} cations. The cations freely diffuse into the interior cavity through the 0.8 nm-sized pores at the fivefold axis of the VNP. Lanthanides have natural affinity for RNA (Feig *et al.*, 1999; Li *et al.*, 2004; Mundoma & Greenbaum, 2002), and inside the VNP the lanthanides form a complex with the encapsidated RNA molecules. The particles were extensively purified via dialysis, gradient ultracentrifugation, and ultrapelleting. Particles remained intact and the lanthanides were stably bound within the interior of the particle. Around 80 ± 20 Gd^{3+} or Tb^{3+} molecules can be encapsulated within CPMV particles (Prasuhn *et al.*, 2007; Singh *et al.*, 2007).

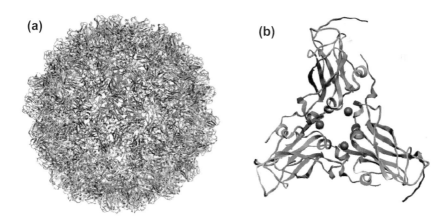

Figure 5.1 Ribbon diagram of the *Cowpea chlorotic mottle virus* (CCMV; pdb 1cwp). (a) The assembled 180 subunits capsid with the asymmetric unit shown in green. (b) Three subunits form the asymmetric unit with putative Ca^{2+} binding site (residues E81, Q85, E148, Q149, and D153 in red, and Trp 55 in blue) coordinating the Ca ions (purple). Figures were provided by courtesy of Prof. Trevor Douglas (Montana State University, MT).

The lanthanides Gd^{3+} and Tb^{3+} can also be bound to the interface of adjacent coat proteins of the CCMV capsid (Allen *et al.*, 2005; Basu *et al.*, 2003). CCMV offers 180 intrinsic metal-binding sites at the threefold axis; under physiological conditions, Ca^{2+} ions are bound to these sites. Coordination of bivalent Ca^{2+} ions involves carboxylic oxygens of three carboxylic acids (Asp/Glu) and carbonyl oxygens of two Gln residues contributed from two adjacent coat protein subunits (Fig. 5.1). Removal of the bivalent Ca^{2+} ions at near-neutral pH causes structural transitions into the swollen, open conformation. This is a result of electrostatic repulsion between the negatively charged groups involved in metal binding (Johnson & Speir, 1997; Speir *et al.*, 1995). Ca^{2+} cations can be replaced with Gd^{3+} or Tb^{3+} cations; binding results in transition into the non-swollen, closed conformation. Up to 180 lanthanides can be bound, and the interaction was found to be stable under physiological conditions (Allen *et al.*, 2005; Basu *et al.*, 2003).

The materials properties of VNP–lanthanide complexes have been evaluated, and it is implied that these formulations would be excellent candidates for further development as contrast agents for magnetic resonance imaging (MRI) (discussed in Chapter 8).

Using a similar approach, small molecules such as fluorescent dyes and drugs have been infused and stably entrapped in particles formed by the plant virus RCNMV (Loo *et al.*, 2008). Like CCMV, RCNMV particles allow pH- and metal ion-dependent reversible gating (Sherman *et al.*, 2006). In the open conformation and in presence of the RNA molecules, small positively charged molecules can freely diffuse into the interior cavity of the particles, where they bind to the negatively charged viral nucleic acids via electrostatic interactions. It was shown that pore opening is required to achieve entrapment, and once entrapped the molecules are stably bound. Pore re-opening does not release the trapped materials. A study reported that around 80 copies of the positively charged fluorescent dye rhodamine and up to 4300 ± 1300 molecules of the cytotoxic drug doxorubicin could be infused and bound within the RCNMV capsid (Loo *et al.*, 2008). The high loading with doxorubicin can be explained by (i) the positive charge of the drug and (ii) its high intrinsic binding affinity for nucleic acids (Hande, 1998). Fluorescent dye-labeled and drug-loaded particles may find further application in biomedical imaging and drug delivery (discussed in Chapter 8).

5.1.1 Triggering Encapsidation of Materials Using RNA Operators

The translational repression (TR) operator of the bacteriophage MS2 has been exploited to facilitate loading of MS2 with drugs (Brown *et al.*, 2002;

Wu *et al.*, 1995). The TR RNA stem loop is a regulatory element controlling protein expression. When assembled MS2 VLPs, in solution or crystalline, are soaked in a solution containing the TR operator, the operator diffuses into the interior of the capsid through 1.4 nm-sized pores in viral capsid. Within the capsid, one TR operator binds to each coat protein dimer; hence, 90 TR operators can be bound within an intact MS2 VLP (MS2 has $T = 3$ symmetry and consists of 180 identical copies of a coat protein).

The TR stem loop can be chemically modified, and studies have shown that small therapeutic compounds such as the plant toxin ricin A chain or 5-fluorouridine can be covalently attached to the operator. Soaking VLPs in a solution containing the drug-modified TR operators led to encapsulation of the drugs within the VLPs. *In vitro* studies using targeted drug-containing MS2 VLPs confirmed specific delivery of the drug inherent with successful cell killing of target cells (discussed in detail in Chapter 8) (Brown *et al.*, 2002; Wu *et al.*, 1995).

5.2 ENCAPSULATION OF MATERIALS DURING PARTICLE SELF-ASSEMBLY

Besides using the gating mechanisms of viruses, such as CCMV or RCNMV, their well-understood *in vitro* self-assembly mechanisms have also been exploited for selective encapsulation of artificial nucleic acids or negatively charged polymers; these techniques were pioneered by Bancroft *et al.* (1969). *In vivo* self-assembly of VNPs is initiated by subtle electrostatic interactions between the RNA and the protein subunits; the nucleic acid assists and stabilizes particle formation. This basic principle can be exploited by mixing coat protein monomers with negatively charged polymers and dialysis against the appropriate assembly buffer. Capsid assembly is initiated via electrostatic interactions of the coat proteins with the negatively charged polymers and results in VLPs encapsulating the artificial cargo (Bancroft *et al.*, 1969; Hu *et al.*, 2008; Sikkema *et al.*, 2007).

In the case of CCMV, assembled VLPs of differing sizes were obtained by mixing the coat protein monomers with different-sized polymers. Polystyrene sulfonate (PSS) of molecular weights ranging from 400 kDa to 3.4 MDa was mixed with CCMV coat proteins and VLPs were re-assembled. Two distinct sizes of CCMV VLPs containing PSS polymers were formed: 22 nm-sized particles with *pseudo* $T = 2$ symmetry were obtained when adding polymers with a molecular weight lower than 2 MDa; 27 nm-sized VLPs resembling $T = 3$ symmetry were formed when polymers of 2 MDa or higher masses were mixed with the coat protein subunits (Fig. 5.2) (Hu *et al.*, 2008).

Another platform that has been exploited for the encapsulation of foreign cargo is *Hibiscus chlorotic ringspot virus* (HCRSV). Similar to the CCMV self-assembly approach, coat protein monomers of HCRSV were mixed with negatively charged polymers, such as polystyrenesulfonic acid (PSA) and polyacrylic acid (PAA). Re-assembly yielded VLPs encapsulating the

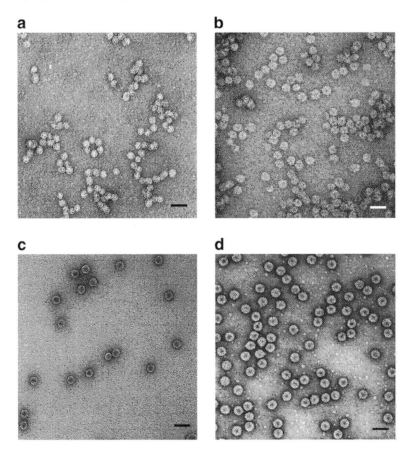

Figure 5.2 Transmission electron micrographs of CCMV capsids formed in self-assembly reactions. Samples were stained with 2% uranyl acetate. (a) VLPs formed with 700-kDa polystyrene sulfonate (PSS). The mean capsid size for VLPs is 22 nm. (b) VLPs formed with 3.4 MDa PSS. The mean capsid size is 27 nm. (c) Empty CCMV capsids formed by dialysis of CP in buffer with high salt and low pH. The dark core in the center indicates the penetration of stain into "void" (aqueous solution) space, which is notably absent in the interiors of VLPs filled with PSS. (d) wt CCMV capsids in virus suspension buffer. Scale bars are 50 nm. Reproduced with permission from Hu, Y., Zandi, R., Anavitarte, A., Knobler, C. M., and Gelbart W. M. (2008) Packaging of a polymer by a viral capsid: the interplay between polymer length and capsid size, *Biophys. J.*, **94**(4), 1428–1436.

polymers as artificial cargo (Ren *et al.*, 2006). Attempts to encapsulate neutral molecules such as fluorescein-labeled dextran failed, demonstrating that encapsulation is based on electrostatic interactions between the negatively charged polymers and the positively charged interior of the capsid proteins (Ren *et al.*, 2006).

In a follow-up study, the concept of encapsulating foreign polymers was exploited to encapsulate a complex of PSA and the cytotoxic drug doxorubicin. The positively charged drug was mixed with PSA and the resulting PSA–

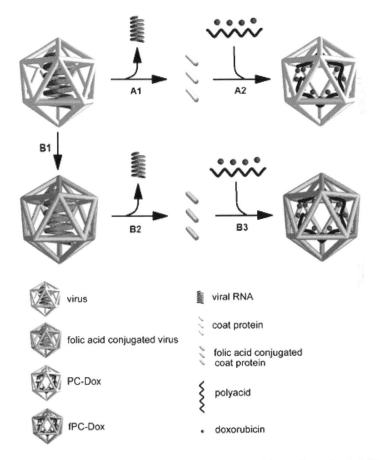

Figure 5.3 Schematic illustration of the preparation of doxorubicin-loaded HCSRV viral protein cage without (PC-Dox) and with folic acid conjugation (fPC-Dox). Steps A1 and B2 are indicative of the removal of viral RNA from the plant virus and purification of coat proteins. Steps A2 and B3 involve the encapsulation of polyacid and doxorubicin during the re-assembly of protein cage. Step B1 refers to the conjugation of folic acid onto the viral protein coat. Reproduced with permission from Ren, Y., Wong, S. M., and Lim, L. Y. (2007) Folic acid-conjugated protein cages of a plant virus: a novel delivery platform for doxorubicin, *Bioconjug. Chem.*, **18**(3), 836–843.

doxorubicin complex added to HCRSV coat protein monomers that were covalently modified with folic acid (a cancer cell-targeting molecules, see Chapter 8). The resulting VLPs combined the therapeutic moieties (doxorubicin, interior) and targeting ligands (folic acid, exterior) and may find applications in targeted drug-delivery approaches (Fig. 5.3) (Ren *et al.*, 2007).

5.1.1.1 Switching from icosahedral to rod-shaped geometry

VNPs are highly dynamic particles, for example, see the above-described reversible swelling mechanisms or recall the complex transitional switches of the bacteriophage HK97 during its maturation process (Section 2.2.8). The coat proteins of TMV are also highly flexible. In experiments reconstituting particles from coat protein monomers, disk- or rod-shaped structures can be obtained (Fraenkel-Conrat & Williams, 1955) (Section 3.5).

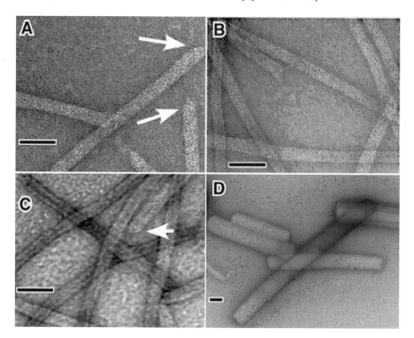

Figure 5.4 Transmission electron microscopy of *in vitro* assembled tubular structures. Uranyl-acetate-stained micrographs of complexes formed at DNA: coat protein dimer ratios of (a) 1 : 1, (b) 7 : 1, (c) 14 : 1, and (d) 28 : 1. CP tube diameter is uniform, about 17 nm, but lengths vary; arrows identify capped ends. DNA condensates (d) have highly variable length and diameter. The scale bars are 50 nm. Reproduced with permission from Mukherjee, S., Pfeifer, C. M., Johnson, J. M., Liu, J., and Zlotnick, A. (2006) Redirecting the coat protein of a spherical virus to assemble into tubular nanostructures, *J. Am. Chem. Soc.*, **128**(8), 2538–2539.

BMV and CCMV particles can be obtained in $T = 1$, *pseudo* $T = 2$, and $T = 3$ symmetry. Furthermore, the coat proteins can be assembled into sheets, stacks, and tubular structures (Bancroft *et al.*, 1967; Krol *et al.*, 1999; Larson *et al.*, 2005; Lucas *et al.*, 2001; Pfeiffer *et al.*, 1976; Pfeiffer & Hirth, 1974) (see Section 3.5). Tubular structures, for example, can be obtained through self-assembly of CCMV coat proteins in the presence of artificial DNA molecules (Mukherjee *et al.*, 2006). Double-stranded DNA molecules were mixed with CCMV coat protein monomers at varying DNA:coat protein ratios and exposed to self-assembly conditions. Tubular structures with a diameter of 17 nm but varying length were obtained (Fig. 5.4) (Mukherjee *et al.*, 2006). The DNA:coat protein ratio governs the length of the tubes. The tubes of lengths ranging from 200 nm to 1 μm were obtained.

5.2.1 Encapsulation of Synthetic Nanoparticles via Templating

Metallic, magnetic, or semiconductor nanoparticles find many applications in materials and medicine. To make use of synthetic nanoparticles as biosensors for imaging or therapeutic approaches, modification of the material is required to (i) make the material biocompatible and (ii) introduce sites for functionalization. Both of these goals can be achieved by encapsulating the material within a VNP. Two basic principles have been used to achieve *in vitro* self-assembly of VLPs around an artificial core (also referred to as templating). Templating can be achieved by decorating a nanoparticle core with a so-called origin-of-assembly site (OAS) that initiates coat protein monomer binding and promotes self-assembly (Section 5.2.1.1). In the second approach, the synthetic nanoparticle core is coated with negatively charged polymers to mimic the negatively charged natural cargo, the nucleic acids (Section 5.2.1.2). Using these approaches, VLPs have been successfully assembled around a range of nanoparticle cores, such as metallic, magnetic, and semi-conducting nanoparticles.

5.2.1.1 Templated VLP assembly via OAS

Templated assembly using an artificial OAS has been studied using the RCNMV platform. The *in vivo* self-assembly process of RCNMV is well understood. The assembly is initiated and stabilized by an internal protein/RNA cage. Formation of the protein/RNA cage structure begins with specific recognition of a viral RNA sequence by the coat protein. This sequence is referred to as OAS. The OAS is a complex of RNA-1 with a RNA-2 stem loop, also known as the trans-activator (Sit *et al.*, 1998). With the detailed knowledge about the OAS, an artificial OAS was created on a gold nanoparticle. Attachment of a thiol-terminated DNA analog of the RNA-2 stem loop (also referred to as

DNA-2) to a gold nanoparticle facilitated hybridization with RNA-1 resulting in an OAS. The OAS-conjugated gold nanoparticles were mixed with RCNMV coat protein monomers, and templated *in vitro* self-assembly of RCNMV VLPs around the gold core was facilitated (Fig. 5.5) (Loo *et al.*, 2006).

This concept was expanded to achieve encapsulation of magnetic nanoparticles and quantum dots (QDs) (Loo *et al.*, 2006, 2007). Nanoparticles with sizes of 5, 10, and 15 nm could be encapsulated. The size of the core correlated with the VLP diameter (Fig. 5.6 and Table 5.1). Attempts to self-assemble particles around 20 nm gold cores were not successful. The RCNMV capsid has an exterior diameter of 36 nm, and the interior cavity has a diameter of 17 nm (Sherman *et al.*, 2006). Given the flexibility and dynamic properties of viral coat proteins, it is interesting that they do not assemble into larger structures when 20 nm-sized cores are provided.

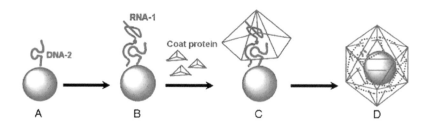

Figure 5.5 Assembly of RCNMV VLPs around a gold nanoparticles via OAS templating. (a) Conjugation of nanoparticle with DNA-2; (b) addition of RNA-1 interacts with DNA-2 to form the functional OAS; (c) the artificial OAS templates the assembly of coat protein; and (d) formation of virus-like particle with nanoparticle encapsidated. Reproduced with permission from Loo, L., Guenther, R. H., Lommel, S. A., and Franzen, S. (2007) Encapsidation of nanoparticles by red clover necrotic mosaic virus, *J. Am. Chem. Soc.*, **129**(36), 11111–11117.

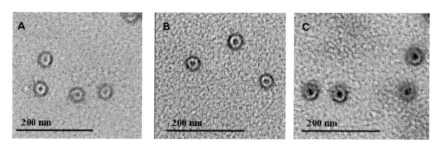

Figure 5.6. Negatively stained transmission electron microscopy images of VLP encapsidated (a) 4 nm $CoFe_2O_4$, (b) 10 nm $CoFe_2O_4$, and (f) 15 nm $CoFe_2O_4$ nanoparticles after purification. Reproduced with permission from Loo, L., Guenther, R. H., Lommel, S. A., and Franzen, S. (2007) Encapsidation of nanoparticles by red clover necrotic mosaic virus, *J. Am. Chem. Soc.*, **129**(36), 11111–11117.

Table 5.1 Effects of encapsulated nanoparticle diameter on the inner and outer diameters of VLPs as measured by transmission electron microscopy[1]

Nanoparticle diameter (nm)	VLP outer diameter (nm)	VLP inner diameter (nm)
Au: 5.0 ± 0.5	30.0 ± 3.0	6.0 ± 0.5
Au: 10.0 ± 1.0	33.5 ± 3.0	10.0 ± 2.0
Au: 15.0 ± 1.5	34.0 ± 2.0	15.0 ± 2.0
$CoFe_2O_4$: 4.0 ± 1.2	33.1 ± 2.2	4.0 ± 0.5
$CoFe_2O_4$: 10.0 ± 3.0	34.2 ± 1.4	9.8 ± 1.1
$CoFe_2O_4$: 15.0 ± 4.5	34.0 ± 2.0	15.0 ± 2.0
QDs: 4.0 ± 1.0	31.6 ± 3.3	4.0 ± 1.5

[1] Reproduced with permission from Loo, L., Guenther, R. H., Lommel, S. A., and Franzen, S. (2007) Encapsidation of nanoparticles by red clover necrotic mosaic virus, *J. Am. Chem. Soc.*, **129**(36), 11111–11117.

5.2.1.2   Encapsulating surface-coated nanoparticle cores

The previous section described the assembly of coat proteins around artificial cores exploiting natural assembly mechanisms and structures: self-assembly is induced and triggered by presentation of an OAS on the nanoparticle core. The next section focuses on a different mechanism: hybrid VLPs are self-assembled around artificial cores based on electrostatic interactions. Nanoparticles coated with negatively charged polymers are utilized. The negative charge mimics the charge of the natural cargo, the nucleic acid, and drives the self-assembly process. The initial approaches to encapsulate gold nanoparticle cores into VLPs were carried out using BMV. Negatively charged citrate-coated nanoparticles with sizes ranging from 2.5 to 4.5 nm were exposed to BMV coat protein monomers. VLPs containing one or two gold nanoparticle cores were obtained; however, only 2% of the assembled VLPs contained the gold cores (Dragnea *et al.*, 2003).

In the following studies, various nanoparticle surface coatings were explored to mimic RNA nucleation and promote efficient *in vitro* self-assembly of VLPs encapsulating the nanoparticle core. Although the citrate coating introduces a negative charge, yields of hybrid VLPs were poor (Dragnea *et al.*, 2003). Surface coatings such as carboxylate-terminated polyethylene glycol (PEG) polymers, lipid micelles, DNA, and dihydrophilic acids were tested (Chen *et al.*, 2006; Dixit *et al.*, 2006; Huang *et al.*, 2007; Sun *et al.*, 2007). Highly

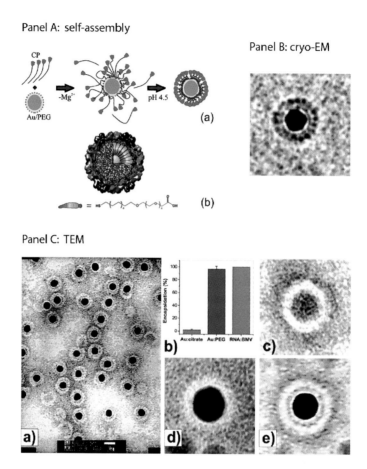

Figure 5.7 Panel A: (a) Proposed mechanism of VLP assembly from coat proteins (CP). First, electrostatic interaction leads to the formation of disordered protein–gold nanoparticle complexes. The second step is a crystallization phase in which the protein–protein interactions lead to the formation of a regular capsid. (b) Schematic depiction of the encapsidated nanoparticle functionalized with carboxyl-terminated TEG chains. Panel B: Cryo-electron micrograph of a single VLP. The regular character of the protein structure coating the 12 nm diameter gold nanoparticle (black disk) is evident. The averages have been obtained by superposition of 10 individual images, in each case. Panel C: (a) Transmission electron micrograph of negatively stained VLPs obtained from functionalized gold nanoparticles (black centers, 12 nm diameter) and BMV capsid protein. (b) Comparison of encapsidation yields for citrate: gold, PEG-gold, and native RNA. Averaged transmission electron micrograph of (c) empty BMV capsid, (d) citrate-coated VLP, and (e) PEG-coated VLP. Reproduced with permission from Chen, C., Daniel, M. C., Quinkert, Z. T., De, M., Stein, B., Bowman, V. D., Chipman, P. R., Rotello, V. M., Kao, C. C., and Dragnea, B. (2006) Nanoparticle-templated assembly of viral protein cages, *Nano Lett.*, **6**(4), 611–615.

efficient encapsulation of gold nanoparticles, magnetic iron oxide nanoparticles, or CdSe/ZnS QDs into BMV VLPs was achieved using DNA or carboxylate-terminated PEG coatings (Chen *et al.*, 2006; Dixit *et al.*, 2006; Huang *et al.*, 2007; Sun *et al.*, 2007). The carboxylate-terminated PEG coating was found to be particularly successful with yields of $95 \pm 5\%$ core-containing VLPs (Fig. 5.7) (Chen *et al.*, 2006). The overall negative charge (derived from the terminating carboxylate groups) mimics the negatively charged natural cargo (the nucleic acid), and the hydrophilic PEG polymers provide a scaffold resistant to non-specific interactions of biomolecules.

5.2.1.2.1 *Size of the core determines the size of the assembled hybrid VLP*

Smaller nanoparticle cores typically yield smaller VLPs. For example, self-assembly of VLPs around 4 nm-sized PEG-carboxylate-coated QDs yielded VLPs that resembled *pseudo T* = 2 symmetry. Interestingly, the majority of the VLPs contained two or three QDs (Fig. 5.8) (Dixit *et al.*, 2006).

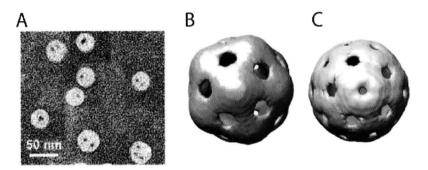

Figure 5.8 (a) Transmission electron microscopy images of VLPs containing carboxylate-terminated PEG-coated QDs. (b, c) Image reconstruction of QD/BMV (b) and T = 2 BMV capsids isolated from plant extracts (c). Reproduced with permission from Dixit, S. K., Goicochea, N. L., Daniel, M. C., Murali, A., Bronstein, L., De, M., Stein, B., Rotello, V. M., Kao, C. C., and Dragnea, B. (2006) Quantum dot encapsulation in viral capsids, *Nano Lett.*, **6**(9), 1993–1999.

The effects of VLP size and symmetry were systematically evaluated using carboxylate-terminated PEG-coated gold cores of 6, 9, 12, 15, and 18 nm (Sun *et al.*, 2007). The most efficient encapsulation occurred with gold cores with a diameter of 12 nm, yielding VLPs similar to wild-type BMV particles and high degree of homogeneity. VLPs assembled from smaller gold cores yielded smaller VLPs, and VLPs assembled from larger gold cores resulted in larger structures (Sun *et al.*, 2007). Two-dimensional crystals were grown for single particle analysis using transmission electron microscopy (TEM) and 3D image reconstruction. VLPs with 6 nm-sized cores yielded particles with $T = 1$ symmetry, VLPs with cores of 9 nm-sized cores gave VLPs resembling *pseudo T = 2* symmetry, and the aforementioned VLPs containing cores with a diameter of 12 nm showed $T = 3$ symmetry (Fig. 5.9) (Sun *et al.*, 2007). Native BMV particles with $T = 1$, *pseudo T = 2*, and $T = 3$ symmetry have also been isolated (Krol *et al.*, 1999; Lucas *et al.*, 2001).

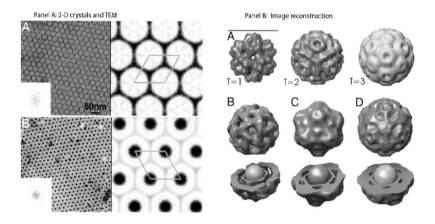

Figure 5.9 Panel A: Negative-stain electron micrographs, Fourier transforms (inserts), and corresponding Fourier projection maps. (a) BMV 2D crystal. The lattice constant is 26 nm (one unit cell is drawn), and the arrangement of the densities suggests a $T = 3$ structure. (b) VLPs containing 12 nm-sized gold cores arranged in a 2D lattice. The lattice constant is 25 nm. Panel B: 3D reconstructions of BMV and VLP using negative stain data. (a) $T = 1$, 2, and 3 models of BMV capsids. The $T = 1$ and pseudo $T = 2$ structures were obtained from the Virus Particle ExplorR (VIPER) database. The $T = 3$ structure is the reconstructed image of BMV in this work (Scale bar, 210 Å.) (b) VLPs containing 6 nm-sized gold cores are characterized by the absence of electron density at the threefold symmetry axes. Its structure and diameter bring it close to a $T = 1$ capsid. (c) The VLPs containing 9 nm-sized gold cores are reminiscent of a pseudo $T = 2$. The presence of electron density at the threefold axes distinguishes it from the VLP structure with 6 nm-sized cores. (d) The shape of the VLPs containing 12 nm-sized gold cores resembles more the spherical shape of BMV, although it still lacks clear evidence of hexameric capsomers. Concentric layering is a characteristic of all VLPs. Reproduced with permission from Sun, J., DuFort, C., Daniel, M. C., Murali, A., Chen, C., Gopinath, K., Stein, B., De, M., Rotello, V. M., Holzenburg, A., Kao, C. C., and Dragnea, B. (2007) Core-controlled polymorphism in virus-like particles, *Proc. Natl. Acad. Sci. USA*, **104**(4), 1354–1359.

The ability to produce particles of distinct sizes makes VNPs valuable building blocks for nanotechnology. Understanding the self-assembly process of VNPs and VLPs is an important goal. Knowledge about the principles of the self-assembly process and an understanding of the factors that govern assembly of particles of a distinct size and symmetry are expected to broaden the applications, versatilities, and possibilities. Using the BMV coat proteins and nanoparticle cores of different sizes, it has been shown that the size of the VLP is governed by the size of the nanoparticle core (Sun *et al.*, 2007). In a more recent study, the relation between changes in the coat protein, such as deletions, and *in vitro* self-assembly around nanoparticles cores has been studied using coat proteins of CCMV. Hybrid nanoparticle-containing VLPs could be assembled even when large portions of the N-terminus of the coat protein (in the native VNP, the N-terminus binds to the nucleic acid and thus plays a critical role in assembly of particles) were deleted (Aniagyei *et al.*, 2009). This again underlines the flexibility of the system.

5.2.1.2.2 *From structural studies to applications*

The potential of nanoparticle-containing VLPs bridges the fields of materials and medicine. The materials properties of gold-core-containing VLP crystals were evaluated for potential application as high-performance metamaterials. A metamaterial is a synthetic material that gains its functional properties, such as optical and electronic properties, from its structure rather than from its composition. The 3D structure of crystalline films of gold-core-containing VLPs has a lattice constant of 28 nm (for VLPs with 12 nm-sized gold cores), and thus might be a good candidate material for optoelectronics. Optical spectroscopy indicated a difference between the optical transmission spectrum of a VLP crystal and that of a dilute solution of VLPs (Fig. 5.10). A double-spectral feature was observed indicating the signature of multipolar coupling between adjacent gold cores leading to plasmonic band formation. The spectrum of the VLP crystal shows two absorption bands, one shifted to the blue and the other one to the red of the surface plasmon peak of the single VLP (Fig. 5.10). Such plasmonic band splittings in metallodielectric materials have been predicted theoretically; however, this has been one of the few examples where it has been shown experimentally (Sun *et al.*, 2007). The surface lattice of the VLP crystal can be varied using different-sized gold nanoparticles as cores. These features suggest that the gold-core-containing VLP crystals are attractive candidates for the development of metamaterials with potential applications in sensors and data storage.

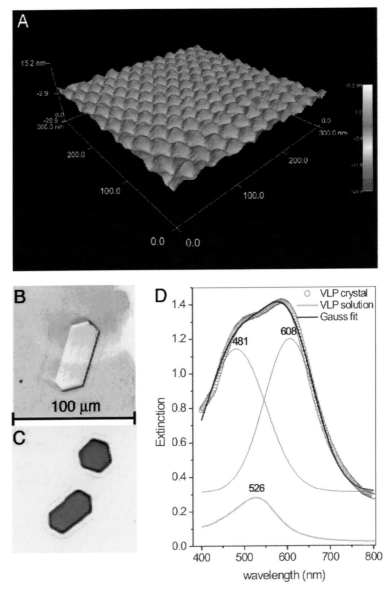

Figure 5.10 VLP crystal properties. (a) AC-mode AFM image of the face of a BMV crystal immersed in its mother liquor. Transmission optical images of BMV (b) and VLP crystals containing 12 nm-sized gold cores (c). (d) Optical spectrum of a 3D VLP crystal is distinct from the single resonance characteristic of dilute VLP crystals containing 12 nm-sized gold cores. Reproduced with permission from Sun, J., DuFort, C., Daniel, M. C., Murali, A., Chen, C., Gopinath, K., Stein, B., De, M., Rotello, V. M., Holzenburg, A., Kao, C. C., and Dragnea, B. (2007) Core-controlled polymorphism in virus-like particles. *Proc. Natl. Acad. Sci. USA*, **104**(4), 1354–1359.

VLPs containing metallic nanoparticles or QDs are also promising candidates for biomedical applications such as imaging and therapy. This will be discussed in detail in Chapter 8; here one example is given to demonstrate the feasibility of these hybrid VLPs for imaging purposes. The above-described assembly protocol can be adapted to other viruses. Successful nanoparticle-templated VLP assembly has been demonstrated using coat proteins from an alphavirus (Goicochea *et al.*, 2007) and *Simian virus* 40 (SV40) (Li *et al.*, 2009), both of which are mammalian viruses. SV40 is a polyomavirus that infects humans and monkeys. Infections are mostly latent, but the virus has the potential for tumor transformation. SV40 is a double-stranded DNA virus. The non-enveloped capsid has icoashedral $T = 7$ symmetry and a diameter of 45 nm (The Universal Virus Database of the International Committee on Taxonomy of Viruses, ICTVdB, http://www.nbci.nlm.nih.gov/ICTVdb).

Common strategies for live imaging of viral trafficking within cells utilize bioconjugation techniques and the covalent attachment of imaging molecules to the exterior surface of the VNP. However, when studying viral entry and intracellular trafficking, exterior surface modifications of virus particles are typically not desired as they can alter and interfere with the natural functions of the capsid such as receptor recognition. Now, it has been demonstrated that QDs with a diameter of 4 nm and coated with mercaptoacetic acid could be encapsulated in SV40 VLPs. SV40 coat protein monomers were mixed with the major coat protein of SV40, the VP1 protein. VLPs with $T = 1$ symmetry and a diameter of 24.1 ± 2.2 nm were self-assembled. The utility of the hybrid material to study viral trafficking in tissue culture has been demonstrated (Fig. 5.11) (Li *et al.*, 2009).

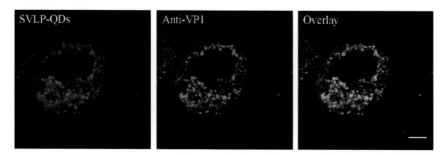

Figure 5.11 SV40-like particles with encapsulated QDs (SVLP-QDs) can "infect" Vero cells. Immunofluorescence using anti-VP1 antibody in the SVLP-QD-loaded (18 h at 37.8 °C) Vero cells. The co-localization between VP1 (green) and QDs (red) confirmed that QDs were carried into Vero cells by VLPs. Scale bars: 10 μm. Reproduced with permission from Li, F., Zhang, Z. P., Peng, J., Cui, Z. Q., Pang, D. W., Li, K., Wei, H. P., Zhou, Y. F., Wen, J. K., and Zhang X. E. (2009) Imaging viral behavior in mammalian cells with self-assembled capsid-quantum-dot hybrid particles, *Small*, **5**(6), 718–726.

5.2.2 Encapsulating Enzymes

Besides encapsulating small chemical or synthetic nanoparticles, the encapsulation of biologically active enzymes into VLPs has also been demonstrated. Coat proteins of CCMV were mixed with the enzyme horseradish peroxidase (HRP), where the conditions were chosen to result in one or zero HRP molecules per assembled VLP. Enzymatic activity was retained (Comellas-Aragones *et al.*, 2007). The enzyme-containing VLPs can be utilized to gain insights into how enzymes work. In biological systems, enzymes are present in confined chemical microenvironments, and one might argue that the enzyme contained within the VLP is thus a good mimic of a particular environment. In addition, studies can be conducted on a single-molecule level.

5.3 CONCLUSIONS

In conclusion, the coat proteins of VNPs can be regarded as remarkably flexible, tunable, and programmable building blocks. Cargo can be entrapped using different strategies, the coat proteins itself can be chemically modified via bioconjugation, and the particles can be self-assembled into different but distinct sizes and shapes. To date, there is no comparable synthetic material that allows this degree of flexibility, underscoring the potential of VNPs for a variety of applications that rely on precise self-assembly.

References

Allen, M., Bulte, J. W., Liepold, L., Basu, G., Zywicke, H. A., Frank, J. A., Young, M., and Douglas, T. (2005) Paramagnetic viral nanoparticles as potential high-relaxivity magnetic resonance contrast agents, *Magn. Reson. Med.*, **54**(4), 807–812.

Aniagyei, S. E., Kennedy, C. J., Stein, B., Willits, D. A., Douglas, T., Young, M. J., De, M., Rotello, V. M., Srisathiyanarayanan, D., Kao, C. C., and Dragnea, B. (2009) Synergistic effects of mutations and nanoparticle templating in the self-assembly of cowpea chlorotic mottle virus capsids, *Nano Lett.*, **9**(1), 393–398.

Bancroft, J. B., Hiebert, E., and Bracker, C. E. (1969) The effects of various polyanions on shell formation of some spherical viruses, *Virology*, **39**(4), 924–930.

Bancroft, J. B., Hills, G. J., and Markham, R. (1967) A study of the self-assembly process in a small spherical virus. Formation of organized structures from protein subunits in vitro, *Virology*, **31**(2), 354–379.

Basu, G., Allen, M., Willits, D., Young, M., and Douglas, T. (2003) Metal binding to cowpea chlorotic mottle virus using terbium(III) fluorescence, *J. Biol. Inorg. Chem.*, **8**(7), 721–725.

Bothner, B., Dong, X. F., Bibbs, L., Johnson, J. E., and Siuzdak, G. (1998) Evidence of viral capsid dynamics using limited proteolysis and mass spectrometry, *J. Biol. Chem.*, **273**(2), 673–676.

Brown, W. L., Mastico, R. A., Wu, M., Heal, K. G., Adams, C. J., Murray, J. B., Simpson, J. C., Lord, J. M., Taylor-Robinson, A. W., and Stockley, P. G. (2002) RNA bacteriophage capsid-mediated drug delivery and epitope presentation, *Intervirology*, **45**(4–6), 371–380.

Chen, C., Daniel, M. C., Quinkert, Z. T., De, M., Stein, B., Bowman, V. D., Chipman, P. R., Rotello, V. M., Kao, C. C., and Dragnea, B. (2006) Nanoparticle-templated assembly of viral protein cages, *Nano Lett.*, **6**(4), 611–615.

Comellas-Aragones, M., Engelkamp, H., Claessen, V. I., Sommerdijk, N. A., Rowan, A. E., Christianen, P. C., Maan, J. C., Verduin, B. J., Cornelissen, J. J., and Nolte, R. J. (2007) A virus-based single-enzyme nanoreactor, *Nat. Nanotechnol.*, **2**(10), 635–639.

Dixit, S. K., Goicochea, N. L., Daniel, M. C., Murali, A., Bronstein, L., De, M., Stein, B., Rotello, V. M., Kao, C. C., and Dragnea, B. (2006) Quantum dot encapsulation in viral capsids, *Nano Lett.*, **6**(9), 1993–1999.

Douglas, T., and Young, M. (1998) Host–guest encapsulation of materials by assembled virus protein cages, *Nature*, **393**, 152–155.

Dragnea, B., Chen, C., Kwak, E. S., Stein, B., and Kao, C. C. (2003) Gold nanoparticles as spectroscopic enhancers for in vitro studies on single viruses, *J. Am. Chem. Soc.*, **125**(21), 6374–6375.

Feig, A. L., Panek, M., Horrocks, W. D., Jr., and Uhlenbeck, O. C. (1999) Probing the binding of Tb(III) and Eu(III) to the hammerhead ribozyme using luminescence spectroscopy, *Chem. Biol.*, **6**(11), 801–810.

Fraenkel-Conrat, H., and Williams, R. C. (1955) Reconstitution of active tobacco mosaic virus from its inactive protein and nucleic acid components, *Proc. Natl Acad. Sci. USA*, **41**(10), 690–698.

Gibbons, M. M., and Klug, W. S. (2007) Mechanical modeling of virus capsids, *J. Mater. Sci.*, **42**, 8995–9004.

Goicochea, N. L., De, M., Rotello, V. M., Mukhopadhyay, S., and Dragnea, B. (2007) Core-like particles of an enveloped animal virus can self-assemble efficiently on artificial templates, *Nano Lett.*, **7**(8), 2281–2290.

Hande, K. R. (1998) Clinical applications of anticancer drugs targeted to topoisomerase II, *Biochim. Biophys. Acta*, **1400**(1–3), 173–184.

Hu, Y., Zandi, R., Anavitarte, A., Knobler, C. M., and Gelbart, W. M. (2008) Packaging of a polymer by a viral capsid: the interplay between polymer length and capsid size, *Biophys. J.*, **94**(4), 1428–1436.

Huang, X., Bronstein, L. M., Retrum, J., Dufort, C., Tsvetkova, I., Aniagyei, S., Stein, B., Stucky, G., McKenna, B., Remmes, N., Baxter, D., Kao, C. C., and Dragnea, B. (2007) Self-assembled virus-like particles with magnetic cores, *Nano Lett.*, **7**(8), 2407–2416.

Johnson, J. E., and Speir, J. A. (1997) Quasi-equivalent viruses: a paradigm for protein assemblies, *J. Mol. Biol.*, **269**(5), 665–675.

Krol, M. A., Olson, N. H., Tate, J., Johnson, J. E., Baker, T. S., and Ahlquist, P. (1999) RNA-controlled polymorphism in the in vivo assembly of 180-subunit and 120-subunit virions from a single capsid protein, *Proc. Natl Acad. Sci. USA*, **96**(24), 13650–13655.

Larson, S. B., Lucas, R. W., and McPherson, A. (2005) Crystallographic structure of the T = 1 particle of brome mosaic virus, *J. Mol. Biol.*, **346**(3), 815–831.

Li, F., Zhang, Z. P., Peng, J., Cui, Z. Q., Pang, D. W., Li, K., Wei, H. P., Zhou, Y. F., Wen, J. K., and Zhang, X. E. (2009) Imaging viral behavior in mammalian cells with self-assembled capsid-quantum-dot hybrid particles, *Small*, **5**(6), 718–726.

Li, S. H., Yuan, W. T., Zhu, C. Q., and Xu, J. G. (2004) Species-differentiable sensing of phosphate-containing anions in neutral aqueous solution based on coordinatively unsaturated lanthanide complex probes, *Anal. Biochem.*, **331**(2), 235–242.

Loo, L., Guenther, R. H., Basnayake, V. R., Lommel, S. A., and Franzen, S. (2006) Controlled encapsidation of gold nanoparticles by a viral protein shell, *J. Am. Chem. Soc.*, **128**(14), 4502–4503.

Loo, L., Guenther, R. H., Lommel, S. A., and Franzen, S. (2007) Encapsidation of nanoparticles by red clover necrotic mosaic virus, *J. Am. Chem. Soc.*, **129**(36), 11111–11117.

Loo, L., Guenther, R. H., Lommel, S. A., and Franzen, S. (2008) Infusion of dye molecules into Red clover necrotic mosaic virus, *Chem. Commun. (Camb)*, (1), 88–90.

Lucas, R. W., Kuznetsov, Y. G., Larson, S. B., and McPherson, A. (2001) Crystallization of Brome mosaic virus and T = 1 Brome mosaic virus particles following a structural transition, *Virology*, **286**(2), 290–303.

Mukherjee, S., Pfeifer, C. M., Johnson, J. M., Liu, J., and Zlotnick, A. (2006) Redirecting the coat protein of a spherical virus to assemble into tubular nanostructures, *J. Am. Chem. Soc.*, **128**(8), 2538–2539.

Mundoma, C., and Greenbaum, N. L. (2002) Sequestering of Eu(III) by a GAAA RNA tetraloop, *J. Am. Chem. Soc.*, **124**(14), 3525–3532.

Pfeiffer, P., Herzog, M., and Hirth, L. (1976) RNA viruses: stabilization of brome mosaic virus, *Philos. Trans. R. Soc. Lond. B Biol. Sci.*, **276**(943), 99–107.

Pfeiffer, P., and Hirth, L. (1974) Aggregation states of brome mosaic virus protein, *Virology*, **61**(1), 160–167.

Prasuhn, D. E., Jr., Yeh, R. M., Obenaus, A., Manchester, M., and Finn, M. G. (2007) Viral MRI contrast agents: coordination of Gd by native virions and attachment of Gd complexes by azide–alkyne cycloaddition, *Chem. Commun.*, (12), 1269–1271.

Ren, Y., Wong, S. M., and Lim, L. Y. (2006) In vitro-reassembled plant virus-like particles for loading of polyacids, *J. Gen. Virol.*, **87**(Pt 9), 2749–2754.

Ren, Y., Wong, S. M., and Lim, L. Y. (2007) Folic acid-conjugated protein cages of a plant virus: a novel delivery platform for doxorubicin, *Bioconjug. Chem.*, **18**(3), 836–843.

Sherman, M. R. H. G., Tama, F., Sit, T. L., Brooks, C. L., Mikhailov, A. M. E. V. O., Baker, T. S., and Lommel, S. A. (2006) *J. Virol.*, **80**, 10395–10406.

Sikkema, F. D., Comellas-Aragones, M., Fokkink, R. G., Verduin, B. J., Cornelissen, J. J., and Nolte, R. J. (2007) Monodisperse polymer-virus hybrid nanoparticles, *Org. Biomol. Chem.*, **5**(1), 54–57.

Singh, P., Prasuhn, D., Yeh, R. M., Destito, G., Rae, C. S., Osborn, K., Finn, M. G., and Manchester, M. (2007) Bio-distribution, toxicity and pathology of cowpea mosaic virus nanoparticles in vivo, *J. Control Release*, **120**(1–2), 41–50.

Sit, T. L., Vaewhongs, A. A., and Lommel, S. A. (1998) RNA-mediated trans-activation of transcription from a viral RNA, *Science*, **281**(5378), 829–832.

Speir, J. A., Munshi, S., Wang, G., Baker, T. S., and Johnson, J. E. (1995) Structures of the native and swollen forms of cowpea chlorotic mottle virus determined by X-ray crystallography and cryo-electron microscopy, *Structure*, **3**(1), 63–78.

Sun, J., DuFort, C., Daniel, M. C., Murali, A., Chen, C., Gopinath, K., Stein, B., De, M., Rotello, V. M., Holzenburg, A., Kao, C. C., and Dragnea, B. (2007) Core-controlled polymorphism in virus-like particles, *Proc. Natl Acad. Sci. USA*, **104**(4), 1354–1359.

Witz, J., and Brown, F. (2001) Structural dynamics, an intrinsic property of viral capsids, *Arch. Virol.*, **146**(12), 2263–2274.

Wu, M., Brown, W. L., and Stockley, P. G. (1995) Cell-specific delivery of bacteriophage-encapsidated ricin A chain, *Bioconjug. Chem.*, **6**(5), 587–595.

Chapter 6

VNPs AS TEMPLATES FOR MATERIALS SYNTHESIS

This chapter gives an overview of the efforts toward utilizing viral nanoparticles (VNPs) as scaffolds or templates for inorganic materials synthesis. Inorganic nanocrystals can be specifically nucleated and grown on the surface of VNPs. The interior cavity of VNPs can also be utilized for spatially controlled synthesis of monodisperse nanocrystals that are trapped within the protein scaffold. The exploitation of naturally occuring organic

Figure 6.1 Elaborate mineral structures produced by diatoms, a major group of eukaryotic algae. Reproduced from http://www.bio.miami.edu/dana/pix/diatoms.jpg.

Viral Nanoparticles: *Tools for Materials Science and Biomedicine*
By Nicole F. Steinmetz and Marianne Manchester
Copyright © 2011 by Pan Stanford Publishing Pte. Ltd.
www.panstanford.com

protein structures, such as VNPs, for the synthesis of man-made materials is referred to as *biotemplating.* The biotemplating approach mimics the process of *biomineralization.* Biomineralization is a natural process by which living organisms produce minerals; proteins orchestrate the formation of complex and inorganic structures with a high level of control (Bauerlein, 2003; Cusack & Freer, 2008; Hildebrand, 2008; Weiner, 2008). Figure 6.1 depicts some of these fascinating structures.

A range of biological templates have been exploited to direct the deposition, assembly, and nucleation of inorganic materials. These include bacterial cell surface layers, also known as S-layers (Sleytr *et al.*, 2004), nucleic acids (Mirkin *et al.*, 1996; Niemeyer, 2001, 2004; Seeman, 2005; Thaxton & Mirkin, 2004), protein cages (see below), VNPs (this chapter), and whole organisms such as prokaryotic and eukaryotic cells (Flenniken *et al.*, 2004; Kowshik *et al.*, 2002; Peelle *et al.*, 2005; Sweeney *et al.*, 2004).

Protein cages are similar to VNPs with regard to their structural organization. They consist of multiple identical copies of protein subunits and form a symmetrical cage structure of repeating subunits with a defined exterior surface and interior cavity. Protein cages, such as the iron storage protein ferritin (which functions to store iron in the form of iron oxide) or heat shock proteins (which function as chaperones that prevent protein denaturation) have been extensively studied and utilized in biomimetic synthesis. The applications of proteins cages in nanotechnology are beyond the scope of this book and the reader is referred to the following reviews (Flenniken *et al.*, 2009; Klem *et al.*, 2005; Uchida *et al.*, 2007; Whyburn *et al.*, 2008; Young *et al.*, 2008).

In the following sections, the utilization of the interior cavity of icosahedral VNPs for materials synthesis (Section 6.1), material deposition on the exterior surface of icosahedral VNPs (Section 6.2), mineralization of the exterior and interior surface of the rod *Tobacco mosaic virus* (TMV) (Section 6.3), and nucleation of nanocrystal wires using genetically engineered M13 phages (Section 6.4) will be discussed. At the end of the chapter, a table is provided that summarizes all the different materials that have been synthesized using VNPs as scaffold (Table 6.1).

6.1 SIZE-CONSTRAINED SYNTHESIS OF INORGANIC MATERIALS WITHIN THE INTERIOR CAVITY OF VNPs

The first efforts to utilize VNPs for materials synthesis were carried out in 1998 by the researchers Douglas and Young (Montana State University,

Bozeman, MT, USA). Empty *Cowpea chlorotic mottle virus* (CCMV) VLPs were employed as size-constrained reaction vessels for the mineralization of the polyoxometalate species paratungstate and decavanadate (Douglas & Young, 1998). The pH-dependent structural transitions of CCMV were coupled to a pH-dependent inorganic oligomerization reaction to load, crystallize, and entrap mineral particles of well-defined size within the VLPs. In brief, aqueous molecular precursors (WO_4^{2-} and VO_3^-) were added to CCMV in the swollen conformation. Next, the pH was lowered to <6.5. By lowering the pH, two effects are induced: (i) a pH-dependent oligomerization of the inorganic species occurs to form large polyoxometalate species ($H_2W_{12}O_{42}^{10-}$ and $V_{10}O_{28}^{6-}$), which were readily crystallized as ammonium salts; and (ii) the structural transition of the CCMV cage from the swollen form to the non-swollen form takes place, trapping the formed material within the viral cage (Fig. 6.2) (Douglas & Young, 1998).

Crystallization occurred only within the capsids; bulk mineralization was not observed. The mineralization reaction is electrostatically driven and induced at the interior surface of the capsid. The exterior and interior capsid surface can be regarded as two chemically distinct environments. Although the exterior of CCMV is not highly charged, the interior is highly positively charged and thus provides an interface for inorganic crystal nucleation. The negatively charged WO_4^{2-} and VO_3^- anions interact with the interior capsid surface via electrostatic interactions; aggregation of the anions then leads

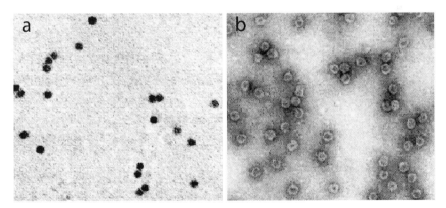

Figure 6.2 Transmission electron micrographs of paratungstate-mineralized CCMV particles after isolation by centrifugation on a sucrose gradient. (a) An unstained sample showing discrete electron dense cores; (b) a negatively stained sample of (a) showing the mineral core surrounded by the intact virus protein cage. Reproduced with permission from Douglas, T., and Young, M. (1998) Host–guest encapsulation of materials by assembled virus protein cages, *Nature*, **393**, 152–155.

to the nucleation and crystallization event (Douglas & Young, 1998). This is similar to the mechanism the iron storage protein ferritin uses to sequester iron: the high charge density on the interior surface of the 12 nm-sized protein cage leads to nucleation of soluble Fe(II), which is subsequently oxidized using O_2 resulting in encapsulated iron oxide (Fe_2O_3) nanoparticles (for a review see Uchida *et al.*, 2007).

These principles can be extended to generate a range of materials within the CCMV cage; the requirement is that the precursors are negatively charged. Materials such as titanium oxide (TiO_2) (Klem *et al.*, 2008) and Prussian blue nanoparticles (de la Escosura *et al.*, 2008) have been synthesized within CCMV. Prussian blue nanoparticles are magnetic materials with potential applications in data storage devices, and the synthesis of monodisperse particles remains challenging. Monodisperse 18 ± 1.7 nm-sized Prussian blue nanoparticles were synthesized within the internal cavity of CCMV. Herein, precursor anions were added to CCMV, and nucleation was initiated by photoreduction (de la Escosura *et al.*, 2008).

The beauty of using VNPs as templates is that one can alter the surface properties and charge of the template using mutagenesis. It was demon-strated that by altering the charge of the interior surface of the CCMV particle from cationic to anionic, the resulting intact particles favored the encapsulation of cationic species (Douglas *et al.*, 2002). A mutant was constructed in which all the basic residues on the N-terminus of the coat proteins were substituted with the negatively charged amino acid Glu (Brumfield *et al.*, 2004). It was found that, because of the electrostatic alterations, the mutant favors strong interaction with ferrous and ferric ions, and that oxidative hydrolysis led to the size-constrained formation of iron oxide (Fe_2O_3) nanoparticles encapsulated in the viral cage (Douglas *et al.*, 2002).

To date, only two VNP platforms have been utilized as containers for nucleation of inorganic materials inside the internal cavity: CCMV and the phage T7 (note: a range of other VNPs have been utilized to encapsulate materials as described in Chapter 5). Empty T7 ghost particles were utilized as nanocontainers and filled with a fluorescent europium complex (Liu *et al.*, 2005) and metallic cobalt (Liu *et al.*, 2006). The synthesis mechanism is the same as described for CCMV, that is, electrostatically driven nucleation of soluble precursors.

6.1.1 Potential Applications of VNPs with Mineralized Cores Lie within the Fields of Data Storage and Medicine

Iron oxide nanoparticles, for example, have been shown to be promising candidates for biomedical imaging such as magnetic resonance imaging

(MRI). The advantage of size-constrained synthesis is, besides monodispersity, that the nanoparticle is encapsidated in a biocompatible protein shell that can be further modified with targeting ligands to facilitate tissue-specific imaging (discussed in Chapter 8).

Magnetic nanoparticles find potential applications in data storage devices, such as magnetic tapes and optical discs. In order to be useful in devices, it is critical to be able to manipulate and organize the particles in a controlled manner onto surfaces. VNPs facilitate highly controlled self-assembly. Design and construction of a range of 1D, 2D, and 3D assemblies can be achieved using a large variety of VNP building blocks and protocols (discussed in Chapter 7). The controlled assembly of VNP–nanoparticle hybrids has led to the development of novel storage devices (Tseng *et al.*, 2006) and battery electrodes (Lee *et al.*, 2009; Nam *et al.*, 2006, 2008), as discussed in the following sections.

6.2 MINERALIZATION OF THE EXTERIOR SURFACE OF ICOSAHEDRAL PARTICLES

Only a few studies have been conducted in which the exterior surface of icosahedral VNPs was exploited for materials synthesis. One hundred and forty nanometer-sized icosahedral $T = 147$ VNPs from the invertebrate virus *Chilo iridescent virus* (CIV) have been utilized as templates to generate gold nanoparticles (Radloff *et al.*, 2005). This was accomplished in a two-step reaction. First, gold nanoparticles (2–5 nm) were seeded on the VNP, where it was found that gold nanoparticles were inherently bound to the capsid in defined patterns. In the second step, gold nanoparticles were used as autocatalytic nucleation sites to promote deposition of gold ions from solution. A complete gold shell was assembled around the CIV template (Fig. 6.3) (Radloff *et al.*, 2005).

The second platform exploited to this end is the 30 nm-sized VNP formed by the $P = 3$ icosahedral *Cowpea mosaic virus* (CPMV). Mineralization with silica (Fig. 6.4) or amorphous FePt was achieved using CPMV mutant particles displaying specific peptide sequences on the exterior surface (Shah *et al.*, 2009; Steinmetz *et al.*, 2009). These peptide sequences allow the specific nucleation of the inorganic material. The discovery, development, and utilization of these peptides will be discussed detail in Section 6.4.

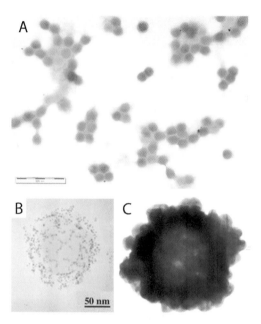

Figure 6.3 *Chilo iridescent virus* (CIV) particles: (a) native CIV particles, (b) dense gold seed coverage. (c) TEM images of gold nanoshell growth on CIV core particles. Reproduced with permission from Radloff, C., Vaia, R. A., Brunton, J., Bouwer, G. T., and Ward, V. K. (2005) Metal nanoshell assembly on a virus bioscaffold, *Nano Lett.,* **5**(6), 1187–1191.

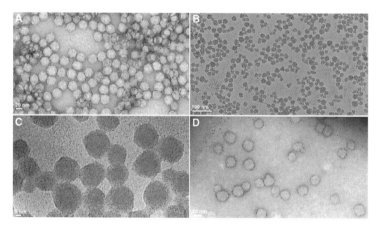

Figure 6.4 Transmission electron micrographs of (a) CPMVsilica-chimera particles before mineralization, stained with uranyl acetate. (b, c) Unstained silicated-CPMVsilica showing dense mineralized particles. (d) Uranyl acetate-stained silicated-CPMVsilica. Reproduced with permission from Steinmetz, N. F., Shah, S. N., Barclay, J. E., Rallapalli, G., Lomonossoff, G. P., and Evans D. J. (2009) Virus-templated silica nanoparticles, *Small*, **5**(7), 813–816.

6.3 MATERIAL DEPOSITION ON THE EXTERIOR AND INTERIOR SURFACE OF TMV

The versatility of TMV as a biotemplate for fabrication of a range of nanotubular inorganic materials via metal deposition has been demonstrated. Nucleation of materials can be achieved on the exterior surface, as well as in the central 4 nm-wide channel. The exterior and interior are chemically distinct, and the different amino acids on the exterior and interior can be regarded as nucleation sites. Amino acid side chains on the interior surface are predominantly carboxylic acids from Asp and Glu; hence, the interior is negatively charged under physiological conditions. The exterior surface

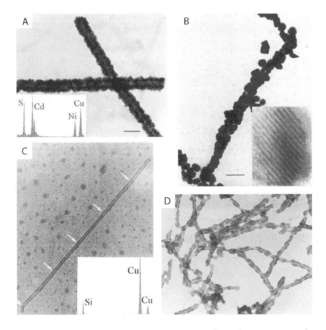

Figure 6.5 Transmission electron microscopy (TEM) images of mineralized TMV structures. (A) CdS-mineralized TMV particles. Scale bar = 50 nm. Inset: corresponding EDX spectrum (Ni and Cu peaks arise from TEM grid and specimen holder, respectively). (b) TEM micrograph showing heavily mineralized PbS/TMV nanotube composite; scale bar = 50 nm. Inset: high-resolution lattice image of a single-domain PbS nanocrystal associated with the TMV external surface. (c) Self-assembled silica/TMV nanotubular superstructure. The white arrows mark the ends of five individual TMV particles, each 300 nm in length, that have aggregated end to end along with a thin external shell of amorphous silica. Inset: corresponding EDX spectrum showing Si peak (Cu peaks are from TEM grid). (d) TEM images of iron-oxide-coated TMV rods. Reproduced with permission from Shenton, W., Douglas, T., Young, M., Stubbs, G., and Mann, S. (1999) Inorganic–organic nanotube composites from template mineralization of *Tobacco mosaic virus*, *Adv. Mater.*, **11**, 253–256.

shows a significant number of Lys and Arg side chains; at around pH 7, the exterior surface exhibits positive charge (Namba & Stubbs, 1986). Thus, the different surface properties of the exterior and interior surface may be exploited to allow spatial and controlled deposition of metals.

In 1999, the first article describing the use of TMV as a biotemplate for mineralization was published (Shenton *et al.*, 1999). The utility of TMV as a template for co-crystallization with PbS and CdS, oxidative hydrolysis of iron oxides, and sol–gel condensation of SiO_2 was demonstrated (Fig. 6.5). The nucleation of PbS and CdS was achieved by incubating TMV with the precursor salts $Pb(NO_3)_2$ or $CdCl_2$, and exposure to H_2S gas. TMV particles were coated with a dense layer of PbS crystallites up to 30 nm in size and about 5 nm-sized CdS nanocrystals, respectively (Fig. 6.5a,b). Exposure of TMV to NaOH and a Fe(II)/Fe(III) solution yielded rods coated with an amorphous iron oxide mineral (Fig. 6.5c). Sol–gel condensation using tetraethoxysilane (TEOS) under acidic conditions resulted in coating TMV with a uniform 3 nm amorphous silica shell. End-to-end alignment of the VNPs was observed, a typical phenomenon of TMV under acidic conditions (Fig. 6.5d) (Shenton *et al.*, 1999).

Noble metals such as platinum, gold, and silver can also be deposited on TMV to form metallic TMV nanotubular composites (Bromley *et al.*, 2008; Dujardin *et al.*, 2003). This is achieved by exposure of TMV to platinum, gold, and silver precursors followed by chemical or photochemical reduction. Platinum and gold coatings were found on the exterior surface whereas silver nanoparticles were incorporated inside the central channel (Fig. 6.6) (Dujardin *et al.*, 2003). The differential nucleation is electrostatically driven and can be explained by the distinct exterior and interior surface charge (positive versus negative). To test this, a TMV mutant with reduced interior anionic charge

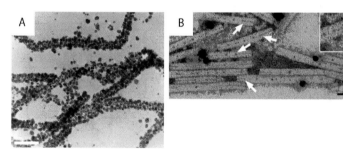

Figure 6.6 Transmission electron micrographs of (a) gold nanoparticles produced on the TMV biotemplate (scale bar is 100 nm) and (b) silver nanoparticles grown inside the hollow channel of TMV. Arrows indicate nanoparticles that prevented the stain (uranyl acetate) from penetrating further in the cavity. One of these is magnified in the inset. Scale bar for the main image is 50 nm. Reproduced with permission from Dujardin, E., Peet, C., Stubbs, G., Culver, J. N., and Mann, S. (2003) Organisation of metallic nanoparticles using *Tobacco mosaic virus*, *Nano Lett.*, **3**, 413–417.

was produced and exposed to silver precursors and reductants. Nucleation of silver occurred preferably on the exterior versus the interior, thus indicating that indeed surface charge and electrostatic interactions are the key players in the mineral deposition mechanism (Dujardin *et al.*, 2003).

A range of methods have been developed that allow coating of TMV with a broad range of inorganic materials (see Table 6.1). These methods include electroless deposition (ELD) and atomic layer deposition (ALD, see below). In ELD, which is also known as autocatalytic or chemical plating, nano-objects are metalized with a metallization bath containing metal ions and a reductant. The redox-reaction leading to the metallization event can only occur on a surface that catalyzes the oxidation of the reductant (the surface can be a noble metal). ELD is an alternative to electroplating where an ionic metal is supplied with electrons to give the desired metal coating.

ELD has been successfully applied to TMV nanostructures, and metals have been plated on the exterior and interior surface. First, TMV is coated and activated with a noble metal, such as Pd, Pt, or Au. The activated TMV particles are then subjected to the metallization bath. The ionic strength and

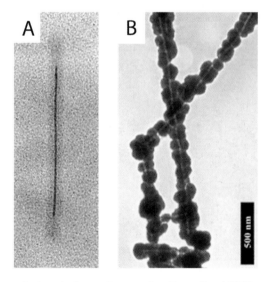

Figure 6.7 Transmission electron micrographs of metalized TMV particles produced by electroless deposition. (a) TMV after Pd(II) activation, followed by electroless deposition of Ni. TMV is filled with a nickel wire with a diameter of approximately 3 nm. Reproduced from Knez M, Bittner AM, Boes F, Wege C, Jeske H, Maiß E, Kern K (2003) Biotemplate synthesis of 3-nm nickel and cobalt nanowires. *Nano Lett.*, **3**(8), 1079–1082. (b) TMV metalized with nickel on the exterior surface. Reproduced with permission from Knez, M., Bittner, A. M., Boes, F., Wege, C., Jeske, H., Maiß, E., and Kern, K. (2004) Spatially selective nucleation of metal clusters on the tobacco mosaic virus, *Adv. Funct. Mater.*, **14**(2), 116–124.

pH conditions have to be carefully selected and adjusted. Using EDL, silver, nickel, cobalt, and copper have been deposited on the exterior or interior of TMV (some examples are given in Fig. 6.7) (Balci *et al.*, 2006; Knez *et al.*, 2002, 2003, 2004).

ALD is mainly used in the microelectronics industry. The ALD process is as follows:

Step 1: The *target* (here TMV) is dried onto a solid support and transferred to the ALD chamber.

Step 2: The target is exposed to a precursor molecule from a gas phase, and the precursors form a layer on the target.

Step 3: The ALD chamber is purged to remove any excess precursor gas.

Step 4: The target is exposed to a second precursor that reacts with the first precursor to give the desired product. The cycle can be repeated until the desired layer thickness is achieved.

ALD has been applied to TMV to yield aluminum oxide and titanium oxide nanostructures. The interior and exterior surfaces of TMV were coated with the mineral; the exterior coating can be removed by ultrasonication to obtain

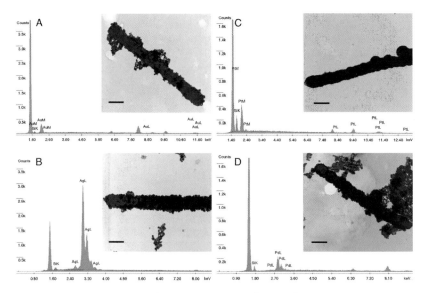

Figure 6.8 Transmission electron micrographs and accompanying energy dispersive X-ray spectra of (a) Au on thick-shell silica-coated TMV template, (b) Ag on thick-shell silica-coated TMV template, (c) Pt on thick-shell silica-coated TMV template, (d) Pd on thick-shell silica-coated TMV template. Al, Fe, and Cu peaks in the EDS spectra are caused by background effects. Scale bar is equal to 100 nm. Reproduced with permission from Royston, E. S., Brown, A. D., Harris, M. T., and Culver, J. N. (2009) Preparation of silica stabilized *Tobacco mosaic virus* templates for the production of metal and layered nanoparticles, *J. Colloid Interface Sci.*, **332**(2), 402–407.

TMV rods with metalized interior only (Knez *et al.*, 2006). TMV disk structures (recall Chapter 3 — Section 3.5) were also obtained. The disk structures contained hollow metal oxide tubes with an exterior diameter of 4 nm and an interior diameter of around 1.0–1.5 nm; these are one of the smallest metal oxide structures generated to date by this method (Knez *et al.*, 2006).

In a different approach, TMV particles coated with a thick silica mineral layer (>20 nm) were used as templates for the metallization reaction. Deposition of metals, such as silver, gold, palladium, and platinum, was accomplished through conventional silica mineralization strategies. Here, mercaptopropyl trimethylsilane (MPS) is used as heterobivalent linker molecule to bridge the silica (silane group of MPS) and the metal (sulfur group of MPS) (Fig. 6.8) (Royston *et al.*, 2009).

The metal nanotubes and nanowires described here are expected to find applications in the field of nanoelectronics; the feasibility of TMV-templated metallic nanomaterials for potential data storage devices has already been reported (Tseng *et al.*, 2006) (see below). For potential industrial applications, it will be of interest to remove the infectious biotemplate prior to embedding into devices. The removal of the TMV template from mineralized structures can be achieved using either a plasma treatment or calcination (thermal degradation). For example, hollow titanium tubes were generated by coating TMV with an ultrathin film of titanium followed by removal of the organic template by oxygen plasma treatment (Fujikawa & Kunitake, 2003). Also, mesoporous inverse silica replicas have been prepared by sol–gel condensation of SiO_2 on TMV followed by calcination (Fowler *et al.*, 2001).

So far, the techniques described in this section have been mostly applied to TMV. In principle, these techniques could be applied to any other VNP. The possibility has been demonstrated; for example, ELD has been applied to CIV-yielding gold-coated icosahedral VNPs (see Section 6.2; Radloff *et al.*, 2005).

6.3.1 TMV Data Storage Device

Platinum-metalized TMV particles served as a building block to fabricate a bio-memory device (Tseng *et al.*, 2006). In brief, TMV particles were coated with Pt nanoparticles via ELD. The Pt nanoparticles were evenly distributed over the surface area of TMV and had an average size of approximately 10 nm (Fig. 6.9a). A functional device was fabricated by embedding the TMV–Pt hybrids in a polyvinyl alcohol (PVA) matrix, which was sandwiched between two aluminum electrodes (Fig. 6.9b). Current–voltage characteristics indicated that the TMV-based device exhibited unique conductance switching behavior.

Initially, the current gradually increases (low-conductance state), upon reaching a turn-on bias, the current increases by more than three orders of magnitude (high-conductance state) (Fig. 6.9b). The low-conductance state can be termed as OFF state whereas the high-conductance state is referred to as ON state. The OFF and ON states can also be denoted as '0' and '1' (to refer to typical data storage terminology) respectively. The system can be switched between OFF and ON (or '0' and '1'). Control experiments were conducted with devices containing Pt or TMV only; however, electrical bistability switching could not be observed, demonstrating that the alignment of the Pt on the TMV bioscaffold is required.

Conductive atomic force microscopy on single TMV–Pt structures confirmed that each individual TMV–Pt wire shows the characteristic switching effect. The device was found to be rewritable and programmable for about 400 cycles. With increasing numbers of cycles, switching becomes more difficult; this could be explained by potential degradation of the TMV structure through Joule heating during programming of the device. Overall, this TMV device can be regarded as a rewritable, non-volatile, 2-bit memory storage device (Tseng *et al.*, 2006). Devices like this can be considered as good candidates for next-generation technologies.

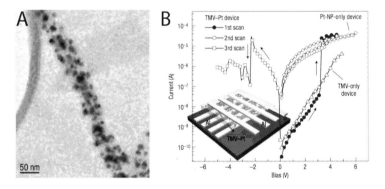

Figure 6.9 (a) Transmission electron microscope (TEM) image of TMV–Pt conjugate. The Pt nanoparticles have an average size of 10 nm and are uniformly attached at the surface of the virus wire. (b) Typical room temperature current–volt (*I–V*) characteristics of the TMV–Pt device. The first bias scan (filled circles) shows that the device switches to the ON state at 3.1 V and stabilizes in the ON state, as confirmed by the second scan (open circles). A reverse scan (squares) indicates that the device turns back to the OFF state at 22.4 V. The *I–V* curves of TMV-only (triangles) and Pt-NP-only (diamonds) devices show no conductance switching. The inset shows a schematic of the device structure with an active layer of dispersed TMV–Pt nanowires. Reproduced with permission from Tseng, R. J., Tsai, C., Ma, L., Ouyang, J., Ozkan, C. S., and Yang, Y. (2006). Digital memory device based on *Tobacco mosaic virus* conjugated with nanoparticles, *Nat. Nano*, **1**, 72.

6.4 BIOTEMPLATING USING GENETICALLY ENGINEERED M13 PHAGES

In the above-described systems, the natural surface properties of VNPs are exploited for metal deposition. The research team led by Belcher (Massachusetts Institute of Technology, MIT; Cambridge, MA, USA) has brought the biotemplating approach to the next level by making use of phage library screening techniques (recall Chapter 2 — Section 2.3.3). The approach uses phage display screening and selection to identify peptide sequences that bind and nucleate specific inorganic materials. These peptides can be genetically engineered to the pIII or pVIII protein and hence be displayed at one end of the phage, or along the virus body. A variety of peptides specific for binding different materials have been identified and utilized on the M13 scaffold (see Table 6.1).

The specificity of the interaction of these peptide sequences with the inorganic counterpart is not fully understood. However, the beauty of the selection process in the phage display system is that it simply allows the discovery of peptides that bind to a desired material, without requiring that the precise mechanism of binding is understood. The peptides are highly selective, and the effective binding motif is rather small; peptides selected are typically from 7-mer libraries (i.e., peptides consisting of seven amino acids).

In a first approach, peptides that bind the crystalline semi-conductors such as GaAs, InP, and Si were selected (Whaley *et al.*, 2000). Figure 6.10 demonstrates the selectivity and specificity of the system, and shows the interaction of a GaAs-specific M13 phage with a semi-conductor heterostructure consisting of GaAs and SiO_2. The phage is exclusively bound to the GaAs surface.

The peptides can be engineered into the pIII protein and thus be displayed on the M13 end structure. When the appropriate metal precursors are added to induce the nucleation and crystal growth, a crystal is formed selectively at the end structure of the phage (Fig. 6.11a) (Lee *et al.*, 2002). Alternatively, the peptides can be fused with the major coat protein pVIII. When the metal precursor are added to such a phage, nucleation and crystal growth occurs throughout the virus body and highly oriented nanowires are synthesized (Fig. 6.11b) (Mao *et al.*, 2003).

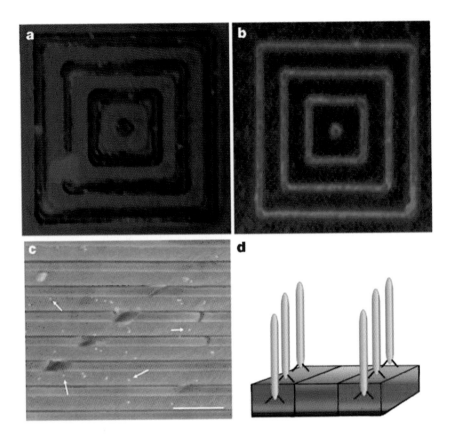

Figure 6.10 Phage recognition of semi-conductor heterostructures. (a, b) Fluorescence images related to GaAs recognition by phage. (a) Control experiment: no phage is present, but primary antibody and streptavidin–tetramethyl rhodamine (TMR) are present. (b) The GaAs clone G12-3 was interacted with a substrate patterned with 1 mm GaAs lines and 4 mm SiO_2 spaces. The phage was then fluorescently labeled with TMR. The G12-3 clone specifically recognized the GaAs and not the SiO_2 surface; scale bar, 4 mm. A diagram of this recognition process is shown in (d), in which phage specifically attach to one semi-conductor rather than another, in a heterostructure. (c) An SEM image of a heterostructure containing alternating layers of GaAs and $Al_{0.98}Ga_{0.02}As$, used to demonstrate that this recognition is element-specific. The cleaved surface was interacted with the GaAs-selective G12-3 phage, and the phage was then tagged with 20 nm gold particles. These nanoparticles (shown arrowed in (c)) are located on GaAs and not AlGaAs layers. Scale bar, 500 nm. Reproduced with permission from Whaley, S. R., English, D. S., Hu, E. L., Barbara, P. F., and Belcher, A. M. (2000) Selection of peptides with semi-conductor binding specificity for directed nanocrystal assembly, *Nature*, **405**(6787), 665–668.

Heterostructures can also be synthesized using dual peptides. This was achieved using phage expressing peptides specific for ZnS and CdS as pVIII fusion. When viruses were incubated with both Zn(II) and Cd(II) in the presence of sulfide anions, the M13 particles nucleated both CdS and ZnS nanocrystals with a stochastic distribution. Both CdS and ZnS nanocrystals appeared on the same phage constructs (Mao *et al.*, 2003). Different peptides can also be introduced at the ends versus virus body. For example, M13 constructs were made, which expressed a gold-binding peptide as pVIII fusion and an anti-streptavidin peptide as pIII fusion. The multifunctional clone was used to assemble a range of metalized structures (Fig. 6.12) (Huang *et al.*, 2005). The gold-binding peptide allows decoration of the phage with gold nanoparticles. The streptavidin peptide allows self-organization, end-to-end or as tripods, via interaction with streptavidin-coated gold nanoparticles (Fig. 6.12) (Huang *et al.*, 2005).

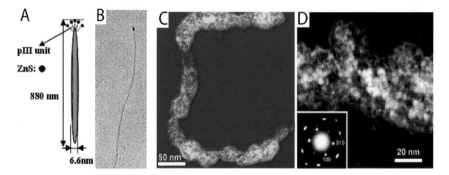

Figure 6.11 (a, b) M13 phage expressing ZnS-specific peptides as pIII fusion. (a) Schematic diagram of the individual M13 phage and ZnS nanocrystals. The pIII peptide unit and the ZnS nanocrystal bound to the phage are not drawn to scale. (b) Transmission electron microscopy (TEM) image of an individual M13 phage (880 nm in length) and ZnS nanocrystals, stained with 2% uranyl acetate. (c, d) M13 phage expressing ZnS-specific peptides as pVIII fusion. (a) High-angle annular detector dark-field scanning TEM (HAADF STEM) image of an individual viral ZnS-virus nanowire. (b) HAADF STEM image of a straight region of a viral nanowire at higher magnification showing the close-packed ZnS nanocrystal morphology. In this image, some areas were brighter as a result of overlapped ZnS nanocrystals. Inset: electron diffraction pattern, taken from the area shown in (b), shows the hexagonal wurtzite ZnS structure. (a) and (b) were reproduced reproduced with permission from Lee, S. W., Mao, C., Flynn, C. E., and Belcher, A. M. (2002) Ordering of quantum dots using genetically engineered viruses. *Science*, **296**(5569), 892–895. Panels c and d were reproduced with permission from Mao, C., Flynn, C. E., Hayhurst, A., Sweeney, R., Qi, J., Georgiou, G., Iverson, B., and Belcher, A. M. (2003) Viral assembly of oriented quantum dot nanowires, *Proc. Natl. Acad. Sci. USA*, **100**(12), 6946–6951.

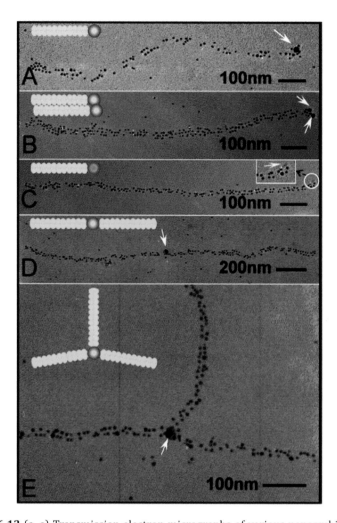

Figure 6.12 (a–e) Transmission electron micrographs of various nanoarchitectures templated by M13. Gold nanoparticles (~5 nm) bind to genetically engineered pVIII proteins along the virus axis and form 1D arrays, whereas a second peptide motif on pIII protein simultaneously binds to streptavidin-coated nanoparticles. Arrows highlight the streptavidin-conjugated gold nanoparticles (~15 nm) and CdSe quantum dots bound on pIII proteins. The insets show the assembly schemes of observed structures. White represents the virus structure, yellow dots represent gold nanoparticles, the green dot represents a CdSe quantum dot, and red represents the streptavidin coating around gold or CdSe particles. (c, inset) The enlarged image of the CdSe quantum dot attached to the end of the virus. Reproduced with permission from Huang, Y., Chiang, C. Y., Lee, S. K., Gao, Y., Hu, E. L., De Yoreo, J., and Belcher, A. M. (2005) Programmable assembly of nanoarchitectures using genetically engineered viruses, *Nano Lett.*, **5**(7), 1429–1434.

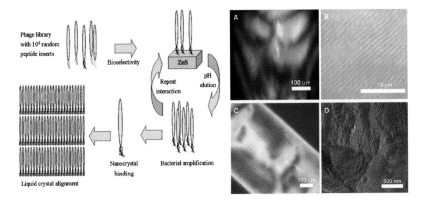

Figure 6.13 Schematic diagram of the process used to generate nanocrystal alignment by the phage display method (left panel). Characterization of the liquid crystalline suspensions of M13 phage–ZnS nanocrystals and cast film. (a) Polarized optical microscopic image of a smectic suspension of M13-ZnS at a concentration of 127 mg/ml. (b) Differential interference contrast imaging brought out dark and bright periodic stripes (~1 μm) that show constructive and destructive interference patterns generated from parallel-aligned smectic layers in the M13-ZnS suspension. (c) The characteristic fingerprint texture of the cholesteric phase of an M13-ZnS suspension (76 mg/ml). (d) Atomic force microscopic micrograph of a cast film from an M13-ZnS suspension (~30 mg/ml) showing close-packed structures of the M13-phage particles. Reproduced with permission from Lee, S. W., Mao, C., Flynn, C. E., and Belcher, A. M. (2002) Ordering of quantum dots using genetically engineered viruses, *Science*, **296**(5569), 892–895.

It has been shown that the metalized M13-based materials can be self-assembled into higher-order structures (Lee *et al.*, 2002, 2006a). For example, M13 phages with terminal ZnS nanocrystals were assembled into films (Fig. 6.13) (Lee *et al.*, 2002). Long-range ordering was accomplished by making use of the intrinsic property of M13 to form liquid crystalline structures (discussed in Chapter 7).

6.4.1 M13-Based Lithium Ion Battery Electrodes

Lithium ion batteries are rechargeable batteries that contain a lithium anode (negative electrode) and a carbon cathode (positive electrode). When in use, lithium ions flow from the lithium anode through polyelectrolyte toward the carbon cathode. This process is referred to as *discharge*. During *charging*, the current is reversed with an external power source; lithium ions diffuse back toward the lithium electrode. Once the original status is restored, the battery is recharged. Energy is stored and released by diffusion of lithium ions between the electrodes.

There has been increased demand in developing novel electrode materials to boost the transport of lithium ions and electrons. Enhanced transport means enhanced storage capacity. Further interest in developing novel electrode materials is driven by lower cost, lower toxicity, and improved thermal and chemical stability.

Initial studies in the Belcher laboratory (MIT; Cambridge, MA, USA) focused on the development of cobalt-oxide-based battery electrodes. Hybrid cobalt oxide nanowires were employed, which were synthesized using the M13 bioscaffold and the above-described peptide-mediated mineralization protocols. Electrochemical evaluation of such wires indicated good discharge and charge capacities; thus, it was proposed that the hybrid would provide a promising system for integration in lithium ion batteries. Two-dimensional film battery electrode assemblies were organized on polyelectrolyte multilayer arrays. Microbatteries were fabricated using a combination of polyelectrolyte-driven assembly and microcontact printing (the assembly techniques are explained in Chapter 7) (Lee *et al.*, 2009; Nam *et al.*, 2006, 2008).

More recently, research has turned toward using iron-phosphate-based materials as components for battery cathodes. A combinatorial genetic approach was used to (i) grow iron phosphate wires using the M13 platform, and (ii) interlink the nanowires with single-walled carbon nanotubes (SWCNTs). To achieve this, two genetic modifications were made: (i) a tetraglutamate peptide (EEEE) was inserted into pVIII (major coat protein) to facilitate synthesis of anhydrous $FePO_4$ (α-$FePO_4$) wires on silver nanoparticle-coated M13 particles; (ii) peptides that specifically bind to SWCNT were inserted in pIII (minor coat protein at one virus end) to allow interlinking and network formation with SWCNTs. Environmentally benign synthesis of anhydrous $FePO_4$ (α-$FePO_4$) nanowires and network formation by mixing with SWCNTs were demonstrated. Carbon nanotubes with their high aspect ratio (length by width ratio) and highly conducting properties were expected to lead to efficient percolating networks (percolation describes the movement or diffusion of fluids, here the lithium ions, through a porous matrix).

Battery electrodes were assembled and tested. The electrodes showed high and stable capacity. The functionality of such M13-based electrodes was demonstrated by powering a light-emitting diode (LED) (Fig. 6.14) (Lee *et al.*, 2009; Nam *et al.*, 2006, 2008). In summary, the nucleation of inorganic materials on the M13 scaffold in combination with its alignment into higher-order heterostructures is an exciting and promising route to novel nanoelectronic devices. Additional data storage and electrode devices are discussed in Chapter 7.

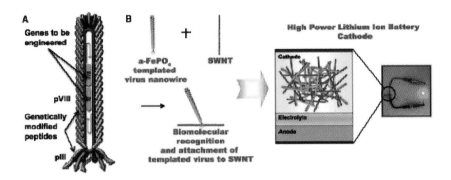

Figure 6.14 Biological toolkits: genetic engineering and biomolecular recognition. (a) A schematic presentation of the multifunctional M13 virus is shown with the proteins genetically engineered in this study. The gene VIII protein (pVIII), a major capsid protein of the virus, is modified to serve as a template for α-FePO$_4$ growth, and the gene III protein (pIII) is further engineered to have a binding affinity for SWNTs. (b) A schematic diagram for fabricating genetically engineered high-power lithium-ion battery cathodes using multifunctional viruses (two-gene system) and a photograph of the battery used to power a green LED. The biomolecular recognition and attachment to conducting SWNT networks make efficient electrical nanoscale wiring to the active nanomaterials, enabling high-power performance. These hybrid materials were assembled as a positive electrode in a lithium-ion battery using lithium metal foil as a negative electrode to power a green LED. Active cathode materials loading was 3.21 mg/cm^2. The 2016 Coin Cell, which is 2 cm in diameter and 1.6 mm in thickness, was used. LED power dissipation was 105 mW. Reproduced with permission from Lee, Y. J., Yi, H., Kim, W. J., Kang, K., Yun, D. S., Strano, M. S., Ceder, G., and Belcher, A. M. (2009) Fabricating genetically engineered high-power lithium-ion batteries using multiple virus genes, *Science*, **324**, 1051.

6.5 OVERVIEW OF INORGANIC MATERIALS SYNTHESIZED ON VNPs

Table 6.1 Overview of the inorganic materials synthesized on VNPs

VNP	Material	Synthesis mechanism	Reference
CCMV (interior)	Molybdate Vanadate	Electrostatically driven synthesis	Douglas and Young (1998)
CCMV (interior)	Iron oxide (γ-FeOOH)	Electrostatically driven synthesis (CCMV mutant with altered surface charge was used as template)	Douglas *et al.* (2002)
CCMV (interior)	Prussian blue	Electrostatically driven synthesis	de la Escosura *et al.* (2008)

CCMV (interior)	Titanium oxide	Electrostatically driven synthesis	Klem *et al.* (2008)
T7 (interior)	Europium	Electrostatically driven synthesis	Liu *et al.* (2005)
T7 (interior)	Cobalt	Electrostatically driven synthesis	Liu *et al.* (2006)
TMV (exterior)	Silica (SiO_2)	Electrostatically driven synthesis	Shenton *et al.* (1999), Royston *et al.* (2006, 2009), Fowler *et al.* (2001)
TMV (exterior)	Iron oxide (Fe_2O_3)	Electrostatically driven synthesis; interaction of metal with metal-binding amino acid side chains	Shenton *et al.* (1999)
TMV (exterior)	CdS	Electrostatically driven synthesis; interaction of metal with metal-binding amino acid side chains	Shenton *et al.* (1999)
TMV (exterior)	PbS	Electrostatically driven synthesis; interaction of metal with metal-binding amino acid side chains	Shenton *et al.* (1999)
TMV (exterior)	Gold	Electrostatically driven synthesis; or silica mineralization chemistry using silica-coated TMV templates	Dujardin *et al.* (2003), Royston *et al.* (2009), Bromley *et al.* (2008)
TMV (interior)	Silver	Electrostatically driven synthesis, or electroless deposition using Pd-, Pt-, or Au-activated TMV particles as templates; or silica mineralization chemistry using silica-coated TMV templates	Dujardin *et al.* (2003), Knez *et al.* (2002), Royston *et al.* (2009)
TMV (exterior)	Platinum	Electrostatically driven synthesis; or electroless deposition; or silica mineralization chemistry using silica-coated TMV templates	Dujardin *et al.* (2003), Lee *et al.* (2006b), Tseng *et al.* (2006), Royston *et al.* (2009)
TMV (exterior)	Palladium	Silica mineralization chemistry using silica-coated TMV templates	Royston *et al.* (2009)

TMV (exterior)	Titanium oxide Aluminum oxide	Atomic layer deposition	Knez *et al.* (2006)
TMV (exterior)	Titanium	Sol–gel process	Fujikawa and Kunitake (2003)
TMV (interior or exterior)	Nickel	Electroless deposition using Pd-, Pt-, or Au-activated TMV particles as templates	Knez *et al.* (2002, 2003, 2004), Balci *et al.* (2006), Royston *et al.* (2008)
TMV (interior or exterior)	Cobalt	Electroless deposition using Pd-, Pt-, or Au-activated TMV particles as templates	Knez *et al.* (2003, 2004), Balci *et al.* (2006), Royston *et al.* (2008)
TMV (exterior)	Copper	Electroless deposition using Pd, Pt, or Au-activated TMV particles as templates	Balci *et al.* (2006)
CIV (exterior)	Gold	Electroless deposition	Radloff *et al.* (2005)
CPMV (exterior)	Silica (SiO_2)	Peptide-mediated synthesis (YSDQPTQSSQRP)[1]	Steinmetz *et al.* (2009)
CPMV (exterior)	FePt	Peptide-mediated synthesis (HNKHLPSTQPLA)	Shah *et al.* (2009)
M13 (exterior)	ZnS	Peptide-mediated synthesis (CNNPMHQNC or VISNHAESSRRL)	Lee *et al.* (2002), Mao *et al.* (2003, 2004)
M13 (exterior)	CdS	Peptide-mediated synthesis (SLTPLTTSHLRS)	Mao *et al.* (2003, 2004)
M13 (exterior)	FePt	Peptide-mediated synthesis (HNKHLPSTQPLA)	Mao *et al.* (2004), Reiss *et al.* (2004)
M13 (exterior)	CoPt	Peptide-mediated synthesis (CNAGDHANC)	Reiss *et al.* (2004), Mao *et al.* (2004)
M13 (exterior)	Co^{2+}	Peptide-mediated synthesis (EPGHDAVP)	Lee *et al.* (2006a)
M13 (exterior)	Gold	Peptide-mediated synthesis (VSGSSPDS)	Nam *et al.* (2006), Souza *et al.* (2006), Huang *et al.* (2005), Khalil *et al.* (2007)

M13 (exterior)	Silver	Peptide-mediated synthesis (EEEE)	Nam *et al.* (2008)
M13 (exterior)	Co$_3$O$_4$	Peptide-mediated synthesis (EEEE)	Nam *et al.* (2006, 2008), Khalil *et al.* (2007)
M13 (exterior)	FePO$_4$	Peptide-mediated synthesis (EEEE)	Lee *et al.* (2009)

[1] Peptide sequences are given in single-letter code.

References

Balci, S., Bittner, A. M., Hahn, K., Scheu, C., Knez, M., Kadri, A., Wege, C., Jeske, H., and Kern, K. (2006) Copper nanowires within the central channel of tobacco mosaic virus pariicles, *Electrochim. Acta*, **51**, 6251–6357.

Bauerlein, E. (2003) Biomineralization of unicellular organisms: an unusual membrane biochemistry for the production of inorganic nano- and microstructures, *Angew. Chem. Int. Ed.*, **42**(6), 614–641.

Bromley, K. M., Patil, A. J., Perriman, A. W., Stubbs, G., and Mann, S. (2008) Preparation of high quality nanowires by tobacco mosaic virus templating of gold nanoparticles, *J. Mater. Chem.*, **18**, 4796–4801.

Brumfield, S., Willits, D., Tang, L., Johnson, J. E., Douglas, T., and Young, M. (2004) Heterologous expression of the modified coat protein of Cowpea chlorotic mottle bromovirus results in the assembly of protein cages with altered architectures and function, *J. Gen. Virol.*, **85**(Pt 4), 1049–1053.

Cusack, M., and Freer, A. (2008) Biomineralization: elemental and organic influence in carbonate systems, *Chem. Rev.*, **108**(11), 4433–4454.

de la Escosura, A., Verwegen, M., Sikkema, F. D., Comellas-Aragones, M., Kirilyuk, A., Rasing, T., Nolte, R. J., and Cornelissen, J. J. (2008) Viral capsids as templates for the production of monodisperse Prussian blue nanoparticles, *Chem. Commun. (Camb)*, (13), 1542–1544.

Douglas, T., Strable, E., and Willits, D. (2002) Protein engineering of a viral cage for constrained material synthesis, *Adv. Mater.*, **14**, 415–418.

Douglas, T., and Young, M. (1998) Host–guest encapsulation of materials by assembled virus protein cages, *Nature*, **393**, 152–155.

Dujardin, E., Peet, C., Stubbs, G., Culver, J. N., and Mann, S. (2003) Organisation of metallic nanoparticles using Tobacco mosaic virus. *Nano Lett.*, **3**, 413–417.

Flenniken, M., Allen, M., and Douglas, T. (2004) Microbe manufacturers of semiconductors, *Chem. Biol.*, **11**(11), 1478–1480.

Flenniken, M. L., Uchida, M., Liepold, L., Kang, S., Young, M. J., and Douglas, T. (2009) A library of protein cage architectures as nanomaterials, *Curr. Top. Microbiol. Immunol.*, **327**, 71–93.

Fowler, C. E., Shenton, W., Stubbs, G., and Mann, S. (2001) Tobacco mosaic virus liquid crystals as templates for the interior design of silicia mesophases and nanoparticles, *Adv. Mater.*, **13**, 1266–1269.

Fujikawa, S., and Kunitake, T. (2003) Surface fabrication of hollow nanoarchitectures of ultrathin titania layers from assembled latex particles and tobacco mosaic viruses as templates, *Langmuir*, **19**, 6545–6552.

Hildebrand, M. (2008) Diatoms, biomineralization processes, and genomics, *Chem. Rev.*, **108**(11), 4855–4874.

Huang, Y., Chiang, C. Y., Lee, S. K., Gao, Y., Hu, E. L., De Yoreo, J., and Belcher, A. M. (2005) Programmable assembly of nanoarchitectures using genetically engineered viruses, *Nano Lett.*, **5**(7), 1429–1434.

Khalil, A. S., Ferrer, J. M., Brau, R. R., Kottmann, S. T., Noren, C. J., Lang, M. J., and Belcher, A. M. (2007) Single M13 bacteriophage tethering and stretching, *Proc. Natl Acad. Sci. USA*, **104**(12), 4892–4897.

Klem, M. T., Young, M., and Douglas, T. (2008) Biomimetic synthesis of -TiO2 inside a viral capsid, *J. Mater. Chem.*, **18**, 3821–3823.

Klem, M. T., Young, M. J., and Douglas, T. (2005) Biomimetic magnetic nanoparticles, *Mater. Today*, **8**, 28–37.

Knez, M., Bittner, A. M., Boes, F., Wege, C., Jeske, H., Maiß, E., and Kern, K. (2003) Biotemplate synthesis of 3-nm nickel and cobalt nanowires, *Nano Lett.*, **3**(8), 1079–1082.

Knez, M., Kadri, A., Wege, C., Gosele, U., Jeske, H., and Nielsch, .K (2006) Atomic layer deposition on biological macromolecules: metal oxide coating of tobacco mosaic virus and ferritin, *Nano Lett.*, **6**(6), 1172–1177.

Knez, M., Sumser, M., Bittner, A. M., Wege, C., Jeske, H., Kooi, S., Burghard, M., and Kern, K. (2002) Electrochemical modification of individual nano-objects, *J. Electroanal. Chem.*, **522**, 70–74.

Knez, M., Sumser, M., Bittner, A. M., Wege, C., Jeske, H., Martin, T. P., and Kern, K. (2004) Spatially selective nucleation of metal clusters on the tobacco mosaic virus, *Adv. Funct. Mater.*, **14**(2), 116–124.

Kowshik, M., Deshmukh, N., Vogel, W., Urban, J., Kulkarni, S. K., and Paknikar, K. M. (2002) Microbial synthesis of semiconductor CdS nanoparticles, their characterization, and their use in the fabrication of an ideal diode, *Biotechnol. Bioeng.*, **78**(5), 583–588.

Lee, S. K., Yun, D. S., and Belcher, A. M. (2006a) Cobalt ion mediated self-assembly of genetically engineered bacteriophage for biomimetic Co–Pt hybrid material,

Biomacromolecules, **7**(1), 14–17.

Lee, S. W., Mao, C., Flynn, C. E., and Belcher, A. M. (2002) Ordering of quantum dots using genetically engineered viruses, *Science*, **296**(5569), 892–895.

Lee, S. Y., Choi, J., Royston, E., Janes, D. B., Culver, J. N., and Harris, M. T. (2006b) Deposition of platinum clusters on surface-modified Tobacco mosaic virus, *J. Nanosci. Nanotechnol.*, **6**(4), 974–981.

Lee, Y. J., Yi, H., Kim, W. J., Kang, K., Yun, D. S., Strano, M. S., Ceder, G., and Belcher, A. M. (2009) Fabricating genetically engineered high-power lithium-ion batteries using multiple virus genes, *Science*, **324**(5930), 1051–1055.

Liu, C. M., Chung, S.-H., Jin, Q., Sutton, A., Yan, F., Hoffmann, A., Kay, B. K., Bader, S. D., Makowski, L., and Chen, L. (2006) Magnetic viruses via nano-capsid templates, *J. Magn. Magn. Mater.*, **302**, 47–51.

Liu, C. M., Jin, Q., Sutton, A., and Chen, L. (2005) A novel fluorescent probe: europium complex hybridized T7 phage, *Bioconjug. Chem.*, **16**(5), 1054–1057.

Mao, C., Flynn, C. E., Hayhurst, A., Sweeney, R., Qi, J., Georgiou, G., Iverson, B., and Belcher, A. M. (2003) Viral assembly of oriented quantum dot nanowires, *Proc. Natl Acad. Sci. USA*, **100**(12), 6946–6951.

Mao, C., Solis, D. J., Reiss, B. D., Kottmann, S. T., Sweeney, R. Y., Hayhurst, A., Georgiou, G., Iverson, B., and Belcher, A. M. (2004) Virus-based toolkit for the directed synthesis of magnetic and semiconducting nanowires, *Science*, **303**(5655), 213–217.

Mirkin, C. A., Letsinger, R. L., Mucic, R. C., and Storhoff, J. J. (1996) A DNA-based method for rationally assembling nanoparticles into macroscopic materials, *Nature*, **382**(6592), 607–609.

Nam, K. T., Kim, D. W., Yoo, P. J., Chiang, C. Y., Meethong, N., Hammond, P. T., Chiang, Y. M., and Belcher, A. M. (2006) Virus-enabled synthesis and assembly of nanowires for lithium ion battery electrodes, *Science*, **312**(5775), 885–888.

Nam, K. T., Wartena, R., Yoo, P. J., Liau, F. W., Lee, Y. J., Chiang, Y. M., Hammond, P. T., and Belcher, A. M. (2008) Stamped microbattery electrodes based on self-assembled M13 viruses, *Proc. Natl Acad. Sci. USA*, **105**(45), 17227–17231.

Namba, K., and Stubbs, G. (1986) Structure of tobacco mosaic virus at 3.6 Å resolution: implications for assembly, *Science*, **231**(4744), 1401–1406.

Niemeyer, C. M. (2001) Nanoparticles, proteins, and nucleic acids: biotechnology meets materials science, *Angew. Chem. Int. Ed.*, **40**, 4128–4158.

Niemeyer, C. M. (2004) DNA-protein nanostructures, in *Nanobiotechnology: Concepts, Applications and Perspectives* (ed. Niemeyer, C. M., and Mirkin, C. A.), Wiley-VCH Verlag GmbH & Co. KGaA, Weinheim, pp. 227–254.

Peelle, B. R., Krauland, E. M., Wittrup, K. D., and Belcher, A. M. (2005) Probing the interface between biomolecules and inorganic materials using yeast surafce display and genetic engineering, *Acta Biomater.*, **1**, 145–154.

Radloff, C., Vaia, R. A., Brunton, J., Bouwer, G. T., and Ward, V. K. (2005) Metal nanoshell assembly on a virus bioscaffold, *Nano Lett.*, **5**(6), 1187–1191.

Reiss, B. D., Mao, C., Solis, D. J., Ryan, K. S., Thomson, T., and Belcher, A. M. (2004) Biological routes to metal alloy ferromagnetic nanostructures, *Nano Lett.*, **4**(6), 1127–1132.

Royston, E., Ghosh, A., Kofinas, P., Harris, M. T., and Culver, J. N. (2008) Self-assembly of virus-structured high surface area nanomaterials and their application as battery electrodes, *Langmuir*, **24**(3), 906–912.

Royston, E., Lee, S. Y., Culver, J. N., and Harris, M. T. (2006) Characterization of silica-coated tobacco mosaic virus, *J. Colloid Interface Sci.*, **298**(2), 706–712.

Royston, E. S., Brown, A. D., Harris, M. T., and Culver, J. N. (2009) Preparation of silica stabilized Tobacco mosaic virus templates for the production of metal and layered nanoparticles, *J. Colloid Interface Sci.*, **332**(2), 402–407.

Seeman, N. C. (2005) Structural DNA nanotechnology: an overview, *Methods Mol. Biol.*, **303**, 143–166.

Shah, S. N., Steinmetz, N. F., Aljabali, A. A. A., Lomonossoff, G. P., and Evans, D. J. (2009) Environmentally benign synthesis of virus-templated, monodisperse, iron-platinum nanoparticles, *Dalton Transaction*, (40), 8479–8480.

Shenton, W., Douglas, T., Young, M., Stubbs, G., and Mann, S. (1999) Inorganic–organic nanotube composites from template mineralization of Tobacco mosaic virus, *Adv. Mater.*, **11**, 253–256.

Sleytr, U. B., Egelseer, E. -M., Pum, D., and Schuster, B. (2004) S-layers, in *Nanobiotechnology: Concepts, Applications and Perspectives* (ed. Niemeyer, C. M., and Mirkin, C. A.), Wiley-VCH Verlag GmbH & Co. KGaA, Weinheim, pp. 77–92.

Souza, G. R., Christianson, D. R., Staquicini, F. I., Ozawa, M. G., Snyder, E. Y., Sidman, R. L., Miller, J. H., Arap, W., and Pasqualini, R. (2006) Networks of gold nanoparticles and bacteriophage as biological sensors and cell-targeting agents, *Proc. Natl Acad. Sci. USA*, **103**, 1215–1220.

Steinmetz, N. F., Shah, S. N., Barclay, J. E., Rallapalli, G., Lomonossoff, G. P., and Evans, D. J. (2009) Virus-templated silica nanoparticles, *Small*, **5**(7), 813–816.

Sweeney, R. Y., Mao, C., Gao, X., Burt, J. L., Belcher, A. M., Georgiou, G., and Iverson, B. L. (2004) Bacterial biosynthesis of cadmium sulfide nanocrystals, *Chem. Biol.*, **11**(11), 1553–1559.

Thaxton, C. S., and Mirkin, C. A. (2004) DNA–gold-nanoparticle conjugates, in *Nanobiotechnology: Concepts, applications and Perspectives* (ed. Niemeyer, C. M., and Mirkin, C. A.), Wiley-VCH Verlag GmbH & Co. KGaA, Weinheim, pp. 288–307.

Tseng, R. J., Tsai, C., Ma, L., Ouyang, J., Ozkan, C. S., and Yang, Y. (2006) Digital memory device based on tobacco mosaic virus conjugated with nanoparticles, *Nat. Nanotechnol.*, **1**, 72–77.

Uchida, M., Klem, M. T., Allen, M., Suci, P., Flenniken, M. L., Gillitzer, E., Varpness, Z., Liepold, L. O., Young, M., and Douglas, T. (2007) Biological containers: protein cages as multifunctional nanoplatforms, *Adv. Mater.*, **19**, 1025–1042.

Weiner, S. (2008) Biomineralization: a structural perspective, *J. Struct. Biol.*, **163**(3), 229–234.

Whaley, S. R., English, D. S., Hu, E. L., Barbara, P. F., and Belcher, A. M. (2000) Selection of peptides with semiconductor binding specificity for directed nanocrystal assembly, *Nature*, **405**(6787), 665–668.

Whyburn, G. P., Li, Y., and Huang, Y. (2008) Protein and protein assembly based material structures, *J. Mater. Chem.*, **18**, 3755–3762.

Young, M., Willits, D., Uchida, M., and Douglas, T. (2008) Plant viruses as biotemplates for materials and their use in nanotechnology, *Annu. Rev. Phytopathol.*, **46**, 361–384.

Chapter 7

PLAYING "NANO-LEGO": VNPs AS BUILDING BLOCKS FOR THE CONSTRUCTION OF MULTI-DIMENSIONAL ARRAYS

The previous chapters discussed the *functionalization* of viral nanoparticles (VNPs) by chemical modification, encapsulation, and mineralization. Now, in order to be useful for functional devices, these functionalized VNPs must be incorporated into other materials, immobilized onto solid supports, or ordered into films or arrays. Films and arrays of nanoparticles are of great importance. The trend, especially in the electronic industry, is to miniaturize devices. A common example is the *iPod Nano*, the smaller version of Apple's *iPod.* In order to manufacture such miniaturized devices, while maintaining efficiency and high-quality performance, smaller building blocks such as wires or data storage chips, are required. Recent developments in nanotechnology, such as viral nanotechnology, have led to the development of materials with high potential for the development of such miniaturized devices. The potential of VNP building blocks for manufacturing battery electrodes, for example, has been shown.

This chapter describes various approaches that allow immobilization of VNPs on surfaces, and strategies in which VNPs have been used as building blocks for the construction of arrays. A range of VNPs have been bound and immobilized onto surfaces. Icosahedral VNPs have been utilized as construction material to fabricate 1D and 2D arrays as well as multi-layered arrays (toward 3D structures). The rods have been used to generate 1D wires, bundles, nanorings, and 2D films.

7.1 THE DIMENSIONS

In mathematics, the term *dimension* refers to the number of coordinates required to locate a point in space. From this viewpoint and for the purposes

Viral Nanoparticles: Tools for Materials Science and Biomedicine
By Nicole F. Steinmetz and Marianne Manchester
Copyright © 2011 by Pan Stanford Publishing Pte. Ltd.
www.panstanford.com

of discussing multi-dimensional structures based on VNPs, an icosahedral VNP could be classified as a 0D structure, where the VNP can be regarded as a simple point in space [However, when considering the surface functional groups (amino acids) on the VNP that allow for the attachment of materials at various angles from the center of the particle, one could also refer to an icosahedron as a 1D structure. In this chapter, icosahedral VNPs are referred to as 0D objects].

A rod-shaped VNP can be regarded as a line and thus has 1D; it takes one coordinate to locate a specific point on the rod. The rods can also be self-assembled into higher-order 1D structures, such as bundles or wires. Icosahedral particles can also be aligned into a 1D structure. Icosahedral and rod-shaped VNPs have also been self-assembled into 2D films or multi-layered 3D arrays (Fig. 7.1).

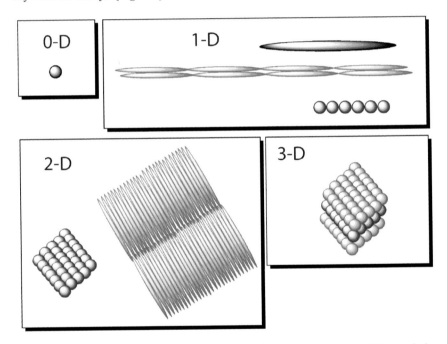

Figure 7.1 Classification of the dimensions. Schematic drawing of VNPs and the structures that can be built with them.

7.2 BUNDLES, WIRES, FIBERS, AND OTHER 1D STRUCTURES BUILT FROM ROD-SHAPED VNPs

In materials science and nanotechnology, there is high demand for robust, long, monodisperse, highly ordered nanofibers and wires. Such highly ordered 1D anisotropic (meaning directionally dependent) structures find applications, especially in nanoelectronic devices. Conducting wires, for example, could be used to interconnect the different electronic components that make up a cell phone or iPod. The rod-shaped VNPs with their high aspect ratio are excellent building blocks for the fabrication of such structures. The aspect ratio of VNPs is defined as length:diameter. M13 and *Tobacco mosaic virus* (TMV) have been extensively studied, and have aspect ratios of approximately 150 and 17, respectively.[1]

7.2.1 Bundles, Wires, and Fibers

Rods can be self-assembled end to end (also referred to as head to tail) and side to side. Structures with high aspect ratios can be built. A collection of rods aligned into a 1D structure is referred to as *bundle*. When these *bundles* are metalized or mineralized (recall Chapter 6), they are referred to as *wires* or *fibers*.

Self-assembly can be spontaneously induced. For example, when exposed to acidic conditions, TMV particles spontaneously self-assemble head to tail. As described in Chapter 6, during the sol–gel silica mineralization process of TMV, which is performed at acidic pH, TMV particles tend to align head to tail and long wires can be formed (see Fig. 6.5) (Shenton *et al.*, 1999). The head-to-tail alignment is most likely a product of complementary hydrophobic interactions between the dipolar ends of the helical structure. In an acidic environment, the head-to-tail alignment is favored as in this conformation repulsion between the carboxylic acid residues at the assembly interface is minimal (Shenton *et al.*, 1999). When such structures are mineralized, a continuous metal coat that is bridging two or several TMV particles is formed, for example, see the silica-coated TMV wires (Fig. 6.5) (Shenton *et al.*, 1999). Continuous wires can also be synthesized in the interior channel of the aligned rods. Figure 7.2 shows a linear TMV aggregate. Straight nickel wires were grown within the central channel that reached lengths of more than 500 nm without any gaps; these results show that linear TMV aggregates have a single uninterrupted central channel and wires can bridge two and potentially multiple virions (Knez *et al.*, 2003).

The head-to-tail assembly of TMV can also be assisted and stabilized by aniline (Niu *et al.*, 2007a,b; Wang *et al.*, 2007). Aniline or phenylamine is an aromatic amine (Fig. 7.3a). Aniline binds to TMV via electrostatic interactions and hydrogen bonding. Polymerization of aniline on the surface of TMV leads to the formation of a thin layer of the polymer on the surface of TMV.

---

[1]M13 has a length of about 1 μm and a diameter of 6.5 nm. The aspect ratio is thus length: diameter = 1000 nm : 6.5 nm = 154. TMV is about 300 nm in length and has a diameter of 18 nm; thus, the aspect is 17.

This *in situ* polymerization stabilizes and fixes the head-to-tail structures formed. Fibers only about 20 nm in width but with lengths up 10 μm can be fabricated using this method (Fig. 7.3b). By varying the pH, bundles of such aligned head-to-tail TMV assemblies can also be fabricated. At neutral pH, aniline prevents lateral organization of TMV and thin fibers consisting of single TMV particles aligned head to tail are assembled. At lower pH, the assembly of bundles is preferred (Fig. 7.3c,d). The formation of bundles can be explained by hydrophobic interactions between the polyaniline at low pH, which promotes parallel alignment of TMV assemblies (Niu *et al.*, 2007a.,b; Wang *et al.*, 2007).

Bundles of the phage M13 have also been fabricated. M13 particles expressing a peptide specific for Co^{2+} ions as pVIII (main coat protein) fusion were utilized for Co^{2+}-induced assembly into phage bundles. CoPt nanoparticles were subsequently synthesized, and a metallic network of phage bundles was created (Lee *et al.*, 2006).

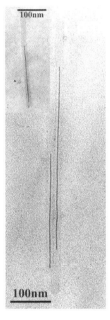

Figure 7.2 Transmission electron microscopy (TEM) image of TMV after Pd(II) activation, followed by electroless deposition of nickel. Two adjacent virion aggregates: the right aggregate comprises three TMV particles in which nickel nanowires were grown. The nickel nanowires are continuous and approximately 600 nm in length. Energy-resolved scanning TEM images proved that the dark wire is indeed composed of nickel. Inset: a single virion is filled with a 200-nm-long wire with a diameter of approximately 3 nm. Reproduced with permission from Knez, M., Bittner, A. M., Boes, F., Wege, C., Jeske, H., Maiß, E., and Kern, K. (2003) Biotemplate synthesis of 3-nm nickel and cobalt nanowires, *Nano Lett.*, **3**(8), 1079–1082.

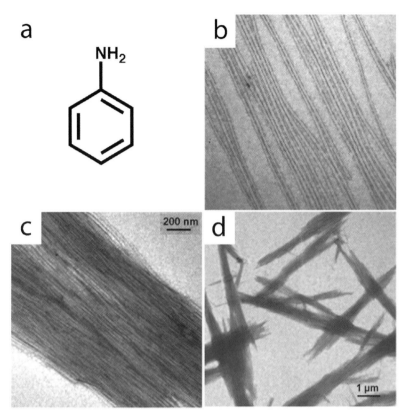

Figure 7.3 (a) Structure of aniline. (b–d) Transmission electron microscopy images of polyaniline/TMV composites synthesized after a 24-h reaction at (b) pH 6.5, (c, d) pH 5.0; scale bar same for (b) and (d). Panels b–d were reproduced with permission from Niu, Z., Bruckman, M. A., Li, S., Lee, L. A., Lee, B., Pingali, S. V., Thiyagarajan, P., and Wang, Q. (2007) Assembly of tobacco mosaic virus into fibrous and macroscopic bundled arrays mediated by surface aniline polymerization, *Langmuir*, **23**(12), 6719–6724.

7.2.2 Nanorings and *Tripods*

Complex 1D assemblies, such as nanorings and tripods have been created using the M13 building block. 1D nanoring structures were constructed by inserting two genetic modifications at each of the opposite ends of M13: a streptavidin-binding peptide was expressed as pIII fusion and a hexa-His tag was inserted in gene encoding pIX. The addition of a heterobifunctional linker — a streptavidin–Ni–NTA complex — to the modified phages resulted in reversible ring formation of the flexible filamentous particle (Fig. 7.4) (Nam *et al.*, 2004).

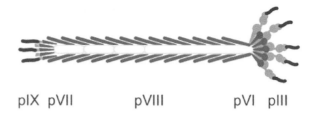

pIX pVII   pVIII   pVI pIII

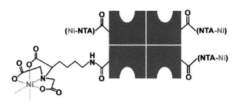

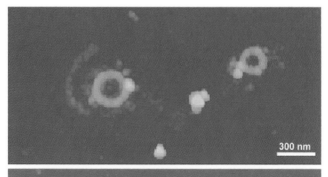

Figure 7.4 Top panel: schematic representation of engineered M13 virus. Hex-His peptide displayed as pIX fusion shown in red, abd anti-streptavidin peptide displayed as pIII fusion shown in blue. (b) Tetrameric streptavidin shown in blue conjugated with four nickel–nitrilotriacetic acid (Ni–NTA) groups. Bottom panel: M13 virus-based ring structures observed by AFM on mica surface. Reproduced with permission from Nam, K. T., Peelle, B. R., Lee, S. W., and Belcher, A. M. (2004) Genetically driven assembly of nanorings based on the M13 virus, *Nano Lett.*, **4**(1), 23–27.

In a different approach, M13 constructs expressing a gold-binding peptide as pVIII fusion and a streptavidin-binding peptide as pIII fusion were used as building blocks. The gold-binding peptide facilitated metallization, whereas the streptavidin-binding peptide allowed self-organization. Up to three phages could be linked to a single streptavidin (streptavidin is a tetrameric protein), and a tripod-like structure was created (see Fig. 6.12) (Huang *et al.*, 2005).

7.3 2D THIN-FILM ASSEMBLIES

Rod-shaped particles have the intrinsic property to form crystalline liquid film assemblies. A liquid crystal is a liquid, in which the molecules (here VNPs) are regularly arrayed in either one or two dimensions. Viral liquid crystals consisting of aligned TMV and M13 particles have been fabricated. Ordering of the VNPs occurs typically in the *nematic* or *smectic* phase; however, ordering into other phases has also been observed. In a *nematic* crystal, the VNPs have long-range orientational order, but do not show any positional order. The VNPs all point in one direction but there is no side-to-side alignment. In contrast, in the *smectic* phase, the VNPs are aligned orientational and also ordered in well-defined layers. One further differentiates a *smectic A* versus a *smectic C* phase. In the *smectic A* phase, the molecules or VNPs are oriented along the main axis; in a *smectic C* phase, the particles are tilted from the main axis (Fig. 7.5).

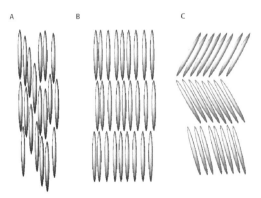

Figure 7.5 Liquid crystalline phases: nematic (a), smectic A (b), and smectic C (c).

Long-range ordered liquid crystalline films of M13 particles have been assembled and studied. When using genetically engineered VNPs that display peptides for biotemplating, metalized films for potential application in electronics can be prepared. When such assemblies are dried, self-supporting films are generated. These films are in the centimeter size range and can be manipulated with forceps (Lee *et al.*, 2002, 2003a,b). Figure 7.6 shows such a film, consisting of M13 particles carrying ZnS quantum dots at the terminal pIII protein. The VNPs are parallel aligned in a zigzag pattern resulting in a smectic C phase liquid crystal (Lee *et al.*, 2002). The thickness of the film can be controlled by virus concentration; the thickness is directly proportional to the film thickness. For example, at a virus concentration ~10 mg/ml, an approximately 13-μm-thick film is assembled. Using a virus concentration of about 1.8 mg/ml, the film thickness lies at about 1.6 μm. By varying the virus concentration, the phase can also be tuned. At a M13 concentration of about 1.8 mg/ml, the VNPs arrange in a nematic phase. At concentrations above, smectic liquid crystals are formed (Lee *et al.*, 2003b). Phase changes (smectic → nematic) can also be controlled by applying a magnetic field (Lee *et al.*, 2002).

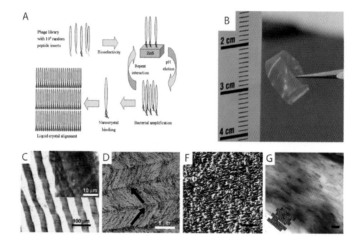

Figure 7.6 (a) Schematic diagram of the process used to generate nanocrystal alignment by the phage display method. (b) Photograph of a M13–ZnS viral film. (c, d) Smectic C liquid crystal. (c) Polarized optical microscopy: birefringent dark and bright band patterns (periodic length 72.8 mm) were observed. These band patterns are optically active, and their patterns reverse depending on the angles between polarizer and analyzer. (d) Atomic force microscopic image of the free surface. Smectic morphologies: the phage forms parallel-aligned herringbone patterns that have almost right angles between the adjacent directors (arrows). (f, g) Nematic liquid crystal. (f) Polarized optical microscopy showing the crooked schlieren dark brush patterns. (g) Atomic force microscopic image of the viral film surface showing the nematic ordering. Reproduced with permission from Lee, S. W., Mao, C., Flynn, C. E., and Belcher, A. M. (2002) Ordering of quantum dots using genetically engineered viruses, *Science*, **296**(5569), 892–895; and Lee, S. W., Woods, B. W., and Belcher, A. M. (2003) Chiral smectic C structures of virus-based films. *Langmuir*, **19**, 1592–1598.

As mentioned above, liquid crystalline films of hybrid mineralized M13 particles can also be constructed (Lee *et al.*, 2002, 2003a). To this end, a phage was engineered that expresses a streptavidin-binding peptide at the terminal pIII protein. The streptavidin-binding peptide provides a unique handle that allows picking up of any material that has been covalently conjugated to streptavidin. In a proof-of-concept experiment, materials such as gold, fluorescein, or phycoerythrin were "picked up" with the phage and then assembled into a liquid crystalline structure (Lee *et al.*, 2003a).

Like M13, TMV also shows liquid crystalline behavior. Convective assembly methods have been used to deposit thin films of highly ordered TMV particles onto surfaces (Fig. 7.7) (Kuncicky *et al.*, 2006; Wargacki *et al.*, 2008). Convective methods are inexpensive, simple, reliable, and versatile. Films in the centimeter size range can be fabricated. It was found that best ordering is achieved on hydrophilic substrates. The film structure and degree

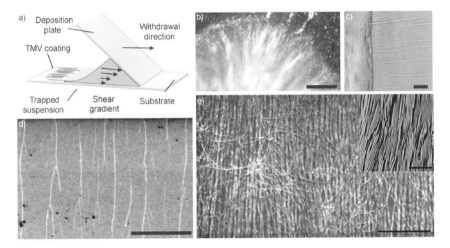

Figure 7.7 (a) Schematic representation of the apparatus for depositing aligned virus-fiber coatings. Typically, virus suspension (20 ml) was entrained between two glass plates held at a fixed angle. The entrained meniscus was dragged across the substrate, and aligned fibers were deposited across the entire surface of the 2.5 × 7.5 cm² substrate. (b) Bright-field image of a freely receding meniscus showing shear-induced alignment in suspension near the three-phase contact line. (c) Phase-contrast optical image of the receding contact line showing TMV-fiber formation on a hydrophobic substrate. Scanning electron microscopic (SEM) image of (d) 5 mg/ml and (e) 40 mg/ml TMV deposited on hydrophilic substrates. Inset: tapping-mode atomic force microscopic (AFM) image of an individual fiber from (e). The scale bars in (b–e) are 500, 100, 5, and 5 mm, respectively, and the scale bar for the inset represents 250 nm. Reproduced with permission from Kuncicky, D. M., Naik, R. R., and Velev, O. D. (2006) Rapid deposition and long-range alignment of nanocoatings and arrays of electrically conductive wires from tobacco mosaic virus, *Small*, **2**(12), 1462–1466.

of ordering are controlled by virus concentration and deposition speed. At low concentration and high deposition speed, 1D fibers are assembled. At higher concentration and slower deposition rate, strong head-to-tail and parallel assembly is induced and thus 2D films are deposited (Wargacki *et al.*, 2008). Metallization of the virus fibers within the film with silver led to conductive arrays of wires with lengths up to many centimeters (Fig. 7.7) (Kuncicky *et al.*, 2006).

Electrospinning and wetspinning techniques can also be applied to generate long-range ordered films of VNPs; this has been demonstrated for M13. In wetspinning, a VNP-containing solution is extruded through a capillary tube to produce fibers. A M13 suspension was extruded through a 20-μm capillary tube into an aqueous glutaraldehyde solution. The glutaraldehyde fixes and cross-links the fibers. Wet-spun fibers were 10–20 μm in diameter and showed nematic ordering of the phages. Electrospinning uses high voltage to draw fibers from a liquid. Electrospun fibers were 100–200 nm thick. To preserve the processing ability and integrity of the VNP during electrospinning, the suspension was blended with the polymer polyvinyl pyrrolidone (PVP). The resulting M13/PVP fibers were continuous and could be transformed into non-woven fabrics (Fig. 7.8) (Lee & Belcher, 2004).

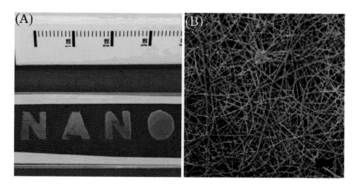

Figure 7.8 Electronspun fiber of M13 virus-blended with polyvinyl pyrolidone. (a) Photograph of non-woven fiber spun through the mask inscribed with the word "NANO", (b) SEM image (scale bar: 1 μm). Reproduced with permission from Lee, S. W., and Belcher, A. M. (2004) Virus-based fabrication of micro- and nanofibers using electrospinning, *Nano Lett.*, **4**(3), 387–390.

7.4 2D ARRAYS OF ICOSAHDRAL VNPs

Icosahedral particles have the intrinsic property to self-assemble into highly ordered 2D or 3D crystals. Self-organized hexagonal 2D monolayers of

VNPs can often be observed on electron microscopic grids. Drying effects and the negative stain (typically uranyl acetate or ammonium molybdate) stabilize the formation of closely packed arrays of VNPs. Examples of self-assembled hexagonal 2D monolayers of *Brome mosaic virus* (BMV) and *Cowpea chlorotic mottle virus* (CCMV) particles are shown in Fig. 7.9 (de la Escosura *et al.*, 2008; Sun *et al.*, 2007).

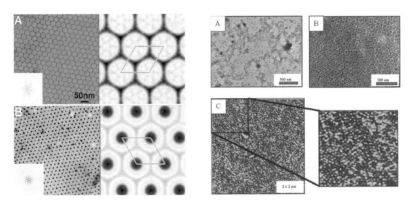

Figure 7.9 Negative-stain transmission electron microscopy (TEM) images hexagonally packed arrays of BMV particles (left panel) and CCMV particles (right panel). Left panel: TEM, Fourier transforms (inserts), and corresponding Fourier projection maps. (a) BMV 2D crystal. The lattice constant is 26 nm (one unit cell is drawn), and the arrangement of the densities suggests a $T = 3$ structure. (b) BMV VLP with encapsulated 12-nm-sized gold nanoparticle, also arranged in a 2D array. The lattice constant is 25 nm. Reproduced with permission from Sun, J., DuFort, C., Daniel, M. C., Murali, A., Chen, C., Gopinath, K., Stein, B., De, M., Rotello, V. M., Holzenburg, A., Kao, C. C., and Dragnea, B. (2007) Core-controlled polymorphism in virus-like particles, *Proc. Natl. Acad. Sci. USA*, **104**(4), 1354–1359. Right panel: TEM micrographs of hexagonally packed arrays of CCMV with Prussian blue core (a) and (b) by TEM on grids stained with ammonium molybdate; (c) by atomic force microscopy on mica. Reproduced with permission from de la Escosura, A., Verwegen, M., Sikkema, F. D., Comellas-Aragones, M., Kirilyuk, A., Rasing, T., Nolte, R. J., and Cornelissen, J. J. (2008) Viral capsids as templates for the production of monodisperse Prussian blue nanoparticles, *Chem. Commun.*, (13), 1542–1544.

7.5 IMMOBILIZATION OF VNPs ON SOLID SUPPORTS

VNPs can be immobilized and bound onto surfaces using a range of chemistries. In the previous section, deposition of VNPs using drying and convective methods, as well as electrospinning and wetspinning, has already been described (see Section 7.2). Such drying methods have also been exploited to form complex and organized patterns of *Cowpea mosaic virus*

(CPMV) particles on surfaces. Concentrated droplets of CPMV were dried onto surfaces such as mica (Fang *et al.*, 2002). Parallel and orthogonal lines with a uniform width and thickness were formed. These finger-like structures can range over hundreds of micrometers. The parallel fingers have a height of 250 nm that equates to nine virus layers. The average width of the fingers was about 600 nm (Fig. 7.10). The formation of patterns was found to be substrate-dependent, where on freshly cleaved mica the particles assembled into parallel and orthogonal lines, whereas on acid-treated mica cross-like patterns were obtained (Fig. 7.10) (Fang *et al.*, 2002).

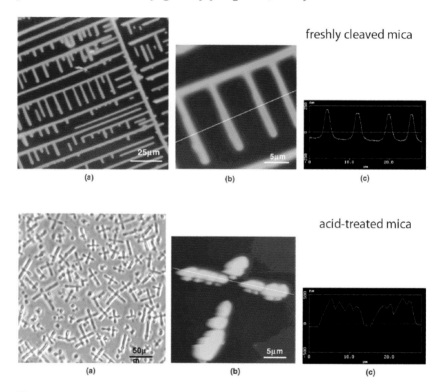

Figure 7.10 Microscopic images of self-assembly from 0.15 mg/ml CPMV solution dried on freshly cleaved mica (top panel) and acid-treated mica (bottom panel). (a) Atomic force microscopic (AFM) image (top panel) and optical microscopy (bottom panel). (b) An enlarged tapping-mode AFM image of (a). (c) Height profile obtained along the line indicated in (b). Reproduced with permission from Fang, J., Soto, C. M., Lin, T., Johnson, J. E., and Ratna, B. R. (2002) Complex pattern formation of *Cowpea mosaic virus* nanoparticles, *Langmuir*, **18**, 308–310.

A range of techniques have been developed that allow capturing VNPs from solution and binding them onto surfaces. In general, one can differentiate between non-specific binding and specifically controlled immobilization. Binding of VNPs onto solid supports can be mediated by (examples and references are given throughout the following sections):

- van der Waals interactions: these are attractive or repulsive forces between molecules based on dipole effects. Polar chemical compounds have dipole moments. The dipole moment is induced by non-uniform distribution of electron density.

- Hydrophobic interactions: hydrophobicity (the word is derived from the Greek language and means "water fearing") is a physical property; hydrophobic molecules are repelled from water and attracted to each other (*birds of a feather flock together*).

- Hydrophilic interactions: hydrophilic molecules *like* water (the word is derived from the Greek language and means "water friendship"). Hydrophilic interactions are mediated via hydrogen bonding that is the interaction between hydrogen atoms with electronegative atoms.

- Covalent bonding: a covalent bond is a bond between two molecules in which pairs of electrons are shared. It is the strongest chemical interaction.

- Electrostatic interactions: these are ionic interactions between oppositely charged molecules. Positively and negatively charged molecules are attracted to each other (*opposites attract*).

- Biospecific interactions: these include the binding of an antibody to its target molecule, or the interaction between the protein streptavidin and its ligand biotin.

Natural binding of TMV on graphite, silicon wafers, mica, glass, and gold surfaces was tested; the strength of the interaction was monitored by atomic force microscopy (Knez *et al.*, 2004). It is important to point out that, in contrast to the above-described studies, here binding is not achieved by drying; rather the particles are captured and bind from the solution phase. Graphite is a hydrophobic and unreactive surface, and TMV was found weakly bound on graphite. Physisorption via van der Waals bonding may explain these interactions. Hydrophilic bonding was observed when TMV was exposed to hydroxyl-containing surfaces such as silicon wafers, mica, or glass. Strongest interaction was observed on acyl-chloride-terminated gold surfaces. Covalent bond formation was confirmed. The acyl chloride termini of the gold surface can form covalent ester links to the viral hydroxyl surface groups (Knez *et al.*, 2004).

7.5.1 Immobilization of VNPs on Gold Surfaces

Cys-added mutants of CCMV, CPMV, and TMV have been utilized to specifically immobilize these VNPs on gold surfaces (Klem *et al.*, 2003; Royston *et al.*, 2008, 2009; Steinmetz *et al.*, 2006). Recall the symmetry-broken Cys-added CCMV particles described in Section 4.3.5. In order to gain control over the assembly of the polyvalent icosahedron CCMV on gold surfaces, particles with single Cys-reactivity were assembled using a solid-state chemical approach (Klem *et al.*, 2003). Non-treated particles that display 180 reactive Cys side chains on the exterior solvent-exposed surface formed

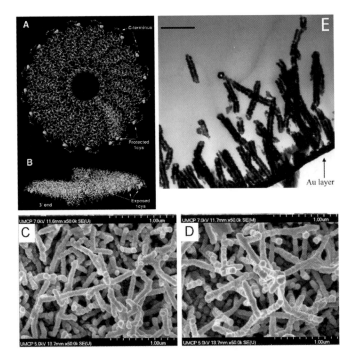

Figure 7.11 Vertically immobilized Cys-added mineralized TMV particles on gold. (a, b) Computer-generated model diagramming the position of the Cys mutations relative to the (a) outer rod surface and (b) the 3′ end. Scanning electron microscopic images showing (c) a nickel-coated gold surface with 1 mg/ml TMV, (d) a cobalt-coated gold surface with 1 mg/ml TMV. (e) Tranmission electron microscopy showing a 70-nm-thick cross-section of nickel-coated TMV attached perpendicular to a gold-coated mica surface. Scale bar is equal to 300 nm. Reproduced with permission from Royston, E., Ghosh, A., Kofinas, P., Harris, M. T., and Culver, J. N. (2008) Self-assembly of virus-structured high surface area nanomaterials and their application as battery electrodes, *Langmuir*, **24**(3), 906–912.

aggregates on surfaces via the formation of disulfide linkages. In contrast, symmetry-broken particles facilitated formation of a controlled monolayer on the substrate (see Fig. 4.14) (Klem *et al.*, 2003).

Vertical patterning of Cys-added TMV particles on gold surfaces has been shown (Fig. 7.11) (Royston *et al.*, 2008, 2009). The favored vertical ordering of the TMV particles was explained by the insertion side of the Cys residue (Fig. 7.11). Although the Cys side chain is surface-exposed, it is somewhat buried on the surface. The Cys is recessed within a groove and partially covered by the carboxyl terminus of the coat protein. This prevents interaction of the virus body with the gold surface. The Cys residue is sufficiently exposed at the 3′ end of the virus, thus allowing vertical attachment of the particles (recall the polar nature of TMV caused by the encapsidated linear RNA molecule; the particles have defined 5′ and 3′ ends; see Section 3.5). The stably immobilized TMV particles could subsequently be mineralized with silica, nickel, or cobalt. Because of the vertical patterning, these metalized nanostructured surfaces offer extremely high surface area and may find applications as electrodes, catalysts, or data storage devices (Royston *et al.*, 2008, 2009).

Indeed, analysis of vertically patterned nickel-coated TMV surfaces indicated the suitability for such a material to function as a battery electrode. Electrode activity was confirmed; TMV-templated electrodes consistently outperformed non-TMV-coated electrodes showing greater capacity. The discharge capacity of nickel-coated TMV-templated battery electrodes was calculated to be on the order of 10^5 mAh/g; for a comparison, commercially available lithium ion battery electrodes have a discharge capacity of 10^4 mAh/g (Royston *et al.*, 2008).

7.5.2 Single VNP *Arrays*

The fabrication of single VNP arrays has the great potential. VNPs immobilized and arranged in a symmetric array offer multiple spatially fixed attachment sites for modification and functionalization, thus allowing precise construction of 2D arrays. VNP arrays offer an extremely large surface area and are thus expected to find applications in manufacturing miniaturized devices. The potential of TMV-based battery electrodes has been demonstrated (see previous section). Nanolithographic and molding techniques have been combined with viral nanotechnology leading to the development of single CPMV, TMV, and M13 VNP arrays (Cheung *et al.*, 2003, 2006; Smith *et al.*, 2003; Suh *et al.*, 2006; Vega *et al.*, 2005).

Nanolithography refers to the fabrication of patterns on the nanometer scale. A range of nanolithographic techniques have been developed; however, only the techniques that have been applied to pattern VNPs are discussed here. For further reading, the following book is suggested: *Nanolithography: A Borderland between STM, EB, IB, and X-ray Lithographies* (Gentili *et al.*, 2008).

To pattern VNPs, *scanning probe nanolithography* (SPN) has been utilized. SPN is a technique using an atomic force microscope (AFM). One differentiates between *dip-pen nanolithography* and *nanografting* (Figs 7.12 and 7.14). In dip-pen nanolithography, the AFM tip is used as a "pen" that is coated with chemical linkers, the "ink". The tip (or "pen") is brought into contact with the substrate that can be regarded as the "paper". The molecules (that is the "ink") are transferred onto the surface (the "paper") using solvent meniscus effects (Fig. 7.12). In nanografting, the AFM tip is used to remove molecules from a coated surface. First, a surface is coated with a chemical (compound A), and then the AFM tip is used in contact mode to draw lines or other structures on the surface by removing compound A from the surface. A different functional chemical compound (compound B) can then be introduced onto the surface (Fig. 7.14).

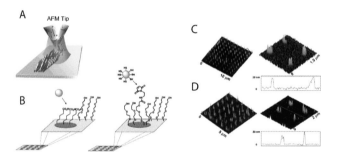

Figure 7.12 (a, b) Schematic of the dip-pen nanolithography process and CPMV binding strategies. (c) Tapping-mode atomic force microscopic (AFM) image, zoom-in image, and height profile of the CPMV wild-type nanoarray (see dotted line). (d) Tapping-mode image, zoom-in image, and height profile (see dotted line) of a CPMV Cys-mutant nanoarray. Reproduced with permission from Smith, J. C., Lee, K., Wang, Q., Finn, M. G., Johnson, J. E., Mrksich, M., and Mirkin, C. A. (2003) Nanopatterning the chemospecific immobilization of cowpea mosaic virus capsid, *Nano Lett.*, **3**, 883–886.

Dip-pen nanolithography has been used to specifically pattern CPMV particles. Gold substrates were patterned with either mercaptohexanoic acid (MHA) or maleimide-terminated polyethylene glycol (PEG) chains. Circular features were "written" on the substrate. The surrounding areas were *passivated* with PEG chains. PEG is a hydrophilic polymer that prevents non-specific adsorption or interaction; hence, the surface area coated with PEG is referred to as *passivated*. VNPs do not bind to the passivated areas.

CPMV particles could be specifically immobilized on the circular nano-sized features by (i) hydrophobic and electrostatic interactions using CPMV wild-type and MHA functional surfaces, and (ii) covalent immobilization was achieved using CPMV Cys-added mutants and maleimide functional templates (Fig. 7.12) (Smith *et al.*, 2003).

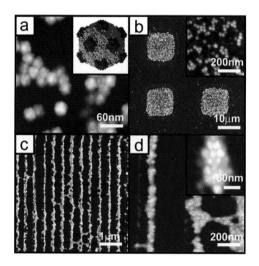

Figure 7.13 Patterned CPMV particles on surfaces. Atomic force microscopic (AFM) height image of CPMV. Inset: model of genetically modified CPMV virus with unique Cys residues (shown as red dots). (b) AFM height image of CPMV virus assembled on micrometer-sized template. Inset: zoom-in section of the functionalized square shown in Fig. 2b. (c) AFM height image of monolayer-thick virions assembled on a parallel line pattern created by nanografting with the chemoselective linkers. (d) Zoom-in section of Fig. 2c. Inset: zoom-in image of another section of the same sample for Fig. 2c. Reproduced with permission from Cheung, C. L., Camarero, J. A., Woods, B. W., Lin, T., Johnson, J. E., and De Yoreo, J. J. (2003) Fabrication of assembled virus nanostructures on templates of chemoselective linkers formed by scanning probe nanolithography, *J. Am. Chem. Soc.*, **125**(23), 6848–6849.

In a similar approach, nanografting was used to pattern a surface that allows capturing and assembling CPMV into lines and squares. Cys-added CPMV particles were covalently immobilized on maleimide-terminated features created by nanografting (Fig. 7.13) (Cheung *et al.*, 2003). This approach has been further extended and the controlled deposition of His-added mutants has also been achieved; CPMV mutants displaying hexa-His sequences were specifically bound onto Ni–NTA features (Cheung *et al.*, 2006). The reaction conditions (virus concentration and buffer composition) were optimized facilitating capturing CPMV in a 1D "monoline", referring to a line consisting of single CPMV particles, where the line width equates to a single CPMV particle (Fig. 7.14) (Cheung *et al.*, 2006).

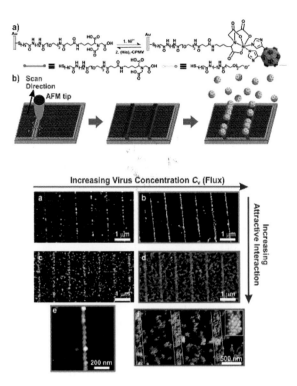

Figure 7.14 Top panel: (a) Schematic of Ni(II) chelation reaction with NTA alkanethiol molecules bound to Au surfaces leading to reversible binding of His-CPMV to NTA-terminated regions and of molecular "inks" used for nanografting process. Short, PEG-terminated alkanethiol resists binding of CPMV (green headgroup); longer, NTA-terminated alkanethiol provides a reversible metal complex linkage to His tags engineered into CPMV (red headgroup). (b) Schematic of nanografting and virus deposition process. Bottom panel: *in situ* AFM images showing evolution of coverage and order during His-CPMV adsorption as the virus flux [i.e., virus concentration (C_v)] and inter-viral interaction (i.e., PEG concentration) are increased. At low flux $(C_v = 0.2$ mg/ml) and 0% PEG (w/w) (a), His-CPMV attaches almost exclusively to Ni–NTA lines, but with poor coverage. At higher flux $(C_v = 2$ mg/ml) and weak interaction (0% PEG) (b), virions still attach to the Ni–NTA lines, but the lines are fully covered. As the inter-viral interaction is increased [$C_v = 0.5$ mg/ml, 1 wt PEG for (c) and $C_v = 2$ mg/ml, 1 wt% PEG for (d)], virus assembly spreads outward from the lines to give stripes that are multiple virions in width and clusters lying between the lines. (e) Higher-resolution AFM image of condition (b) showing the single line of His-CPMV particles. (f) High-resolution image of condition (d) showing a packing geometry similar to that seen for the ordered virus cluster in Fig. 1e. Times at which images were collected are (a) 4, (b) 8, (c) 4, and (d) 6 h. Reproduced with permission from Cheung, C. L., Chung, S. W., Chatterji, A., Lin, T., Johnson, J. E., Hok, S., Perkins, J., De Yoreo, J. J. (2006) Physical controls on directed virus assembly at nanoscale chemical templates, *J. Am. Chem. Soc.*, **128**(33), 10801–10807.

The assembly of nanopatterned arrays of single TMV particles was achieved using a combination of dip-pen nanolithography and coordination chemistry. Precise immobilization and positioning of single VNPs was demonstrated (Fig. 7.15) (Vega *et al.*, 2005). The chemical immobilization approach relies on the ability of metal ions (such as Zn^{2+}) to bridge carboxylates from an MHA-patterned surface with TMV surface carboxylates. In brief, MHA patterns were generated using dip-pen nanolithography and the surrounding surface was passivated with PEG. Coordination with Zn^{2+} was achieved by exposing the surface to a solution of $Zn(NO_3)_2$. In the subsequent step, TMV was introduced. Different feature sizes were tested. The ideal template consisted of 350 × 110 nm-sized features (TMV has dimensions of 300 × 18 nm). Using this template, single 2D arrays of TMV were fabricated (Fig. 7.15). Smaller feature sizes resulted in a less uniform assembly, where not all sites were occupied with VNPs. Larger feature sizes could result in several VNPs binding to one site. Dot templates with features of 350 nm in diameter also facilitated the fabrication of 2D arrays. TMV particles bound to the dot features adopted a curved structure (Fig. 7.15) (Vega *et al.*, 2005).

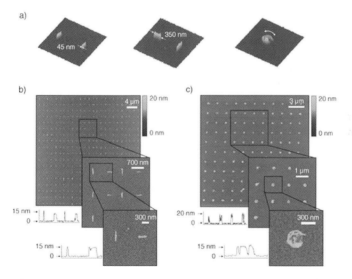

Figure 7.15 AFM tapping-mode images and height profiles of TMV nanoarrays: (a) 3D topographical images of pairs of virus particles within larger arrays: a parallel array (left), a perpendicular array (middle), and dot arrays (right); (b) topographic images and height profiles of a perpendicular array of single virus particles (40 × 40 mm²); (c) topographic image and height profiles of a TMV nanoarray (20 × 20 mm²) formed on an array of MHA dot features (diameter = 350 nm) pre-treated with $Zn(NO_3)_2 \cdot 6H_2O$. Reproduced with permission from Vega, R. A., Maspoch, D., Salaita, K., and Mirkin, C. A. (2005) Nanoarrays of single virus particles, *Angew. Chem. Int. Ed.*, **44**(37), 6013–6015.

Soft lithographic techniques such as *nanomolding* have been applied to design single M13 arrays. In brief, a patterned polydimethylsiloxane (PDMS) stamp was placed on a wet-spin-coated film of PEG polymers. Solvent-assisted capillary effects during the drying process result in the formation of patterns. The pattern is a replica of the stamp features. Nanowells were *molded* into the film (Fig. 7.16). The wells were functionalized with antibodies specific for the M13 pIII coat protein, which facilitated capturing of individual M13 particles within the wells (Fig. 7.16) (Suh *et al.*, 2006).

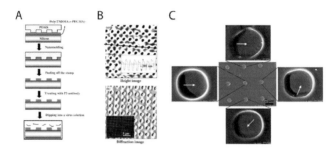

Figure 7.16 (a) Schematic illustration of the experimental procedure. Prior to virus seeding, the patterned surface was treated with an antibody specific for the pIII M13 coat protein to promote adhesion of the virus. (b) Atomic force microscopic images of 700 nm PEG nanowells in (a) height and (b) deflection modes. Scan area was $15 \times 15 \, \mu m^2$. The inset shows the corresponding fluorescent image treated with the P3 antibody and a FITC-labeled secondary antibody. (c) SEM image of an individual virus array on six wells. Four insets show the captured single virus at different locations. The arrows indicate the location of the virus. Reproduced with permission from Suh, K. Y., Khademhosseini, A., Jon, S., and Langer, R. (2006) Direct confinement of individual viruses within polyethylene glycol (PEG) nanowells, *Nano Lett.*, **6**(6), 1196–1201.

7.5.3 Immobilization of VNPs Using Biospecific Interactions

VNPs can be immobilized on surfaces using biospecific interactions. Biospecific interactions were used in the above-described example, in which M13 particles have been captured in nanowells using an antibody specific for the pIII coat protein (Suh *et al.*, 2006). Furthermore, biospecific interactions used to immobilize VNPs include the biotin–streptavidin system, Ni–NTA–hexa-His interactions, and nucleic acid hybridization. The biotin–streptavidin system has been widely used for the construction of multi-layered arrays of VNPs and will be discussed in Section 7.6. Immobilization of His-mutant VNPs that display a hexa-His tag can be achieved via the strong interaction of the hexa-His sequence with Ni–NTA. This strategy has been used for CPMV and $Q\beta$ (Chatterji *et al.*, 2005, 2006; Medintz *et al.*, 2005; Udit *et al.*, 2008) (see also Section 7.5.3).

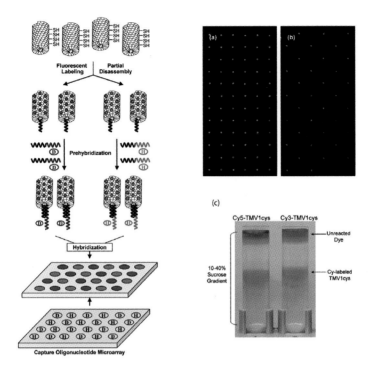

Figure 7.17 TMV microarrays. Left panel: hybridization-based programming of fluorescently labeled and partially disassembled TMV nanotemplates for assembly onto multiple addresses on DNA oligonucleotide microarray platforms. Right panel: patterned assembly of fluorescently labeled and programmed TMV nanotemplates onto oligonucleotide microarrays via hybridization of Cy5-TMV1cys-D and Cy3-TMV1cys-H simultaneously. Cy5 and Cy3 are fluorescent dyes, the TMV particles used were Cys-added mutants referred to as TMV1cys, and D and H denote different oligonucleotide specificities. (a) Alternating pattern of sequences D and H. (b) Alternating pattern of sequences G and H. (c) Sucrose gradient centrifugation of Cy5- and Cy3-labeled TMV nanotemplates. Unreacted dyes remain at the top, whereas labeled TMVs appear in the middle as distinctive bands. Reproduced with permission from Yi, H., Nisar, S., Lee, S. Y., Powers, M. A., Bentley, W. E., Payne, G. F., Ghodssi, R., Rubloff, G. W., Harris, M. T., and Culver, J. N. (2005) Patterned assembly of genetically modified viral nanotemplates via nucleic acid hybridization, *Nano Lett.*, **5**(10), 1931–1936.

TMV has been specifically immobilized on surfaces via nucleic acid hybridization (Tan *et al.*, 2008; Yi *et al.*, 2005, 2007). The polar nature of TMV was utilized. The helical encapsulation of the RNA molecule results in sequence-definable 5′ and 3′ ends. A mild disassembly protocol was used to partially disassemble the protein coat and expose the RNA at the 5′ end.

Spatially oriented assembly of TMV on solid supports was then achieved in a controlled manner via nucleic acid hybridization using complementary oligonucleotides. Using this strategy, TMV particles were immobilized onto electrodes and micropatterned surfaces as well as on micropatterned microcapsules (Tan *et al.*, 2008; Yi *et al.*, 2005, 2007). Using differentially labeled TMV particles and a micropatterned substrate, the construction of a patterned TMV microarray was achieved (Fig. 7.17) (Yi *et al.*, 2007).

7.6 ASSEMBLY OF VNPs AT LIQUID–LIQUID INTERFACES

Besides assembling VNPs onto solid supports, VNPs can also be assembled at liquid–liquid interfaces. A liquid–liquid interface is the phase between two liquids. For example, when oil is immersed in water, oil droplets are formed.

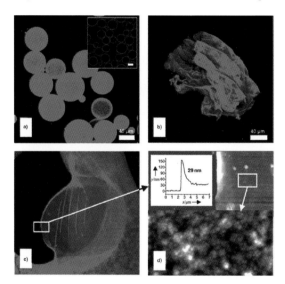

Figure 7.18 Confocal fluorescence microscopic image of CPMV particle assembly after cross-linking with glutaraldehyde. (a) Three-dimensional reconstruction of perfluorodecalin droplets in water that are coated with the virus (inset: cross-sectional view). Excess particles were removed by successive washing with water. (b) Crumpled droplet after complete drying and rehydration with water. (c) Capsule cap after complete drying. (d) The white box shows the area at which the SFM scan was taken, and the lower part shows the height profile on top of the collapsed capsule (image width = 2 mm, Z-range = 30 nm). Reproduced with permission from Russell, J. T., Lin, Y., Boker, A., Su, L., Carl, P., Zettl, H., He, J., Sill, K., Tangirala, R., Emrick, T., Littrell, K., Thiyagarajan, P., Cookson, D., Fery, A., Wang, Q., and Russell, T. P. (2005) Self-assembly and cross-linking of bionanoparticles at liquid-liquid interfaces, *Angew. Chem. Int. Ed.*, **44**(16), 2420–2426.

The interface between the oil droplet and surrounding water is a liquid–liquid interface. CPMV or TMV particles, for example, have been self-assembled at such water–oil interfaces (He *et al.*, 2009; Russell *et al.*, 2005). Figure 7.18 shows a monolayer of fluorescent-labeled CPMV particles self-assembled on perfluorodecalin (oil) droplets in aqueous buffer. CPMV particles segregated at the oil–water interface, thus stabilizing the oil droplets. The CPMV monolayer can be cross-linked using glutaraldehyde or biotinylated CPMV particles in combination with streptavidin (Russell *et al.*, 2005).

7.7 ARRAYS OF VNPs LAYER-BY-LAYER

Assembly of multi-layered arrays toward the assembly of 3D structures consisting of nanoparticles or biomolecules can be achieved via layer-by-layer (LbL) assembly. Multi-layered thin-film assemblies are of growing interest for the development of miniaturized sensors, reactors, and biochips. LbL self-assembly methods have been developed, allowing the deposition of various components in a defined and ordered way. The multi-layered structure is self-assembled one layer at a time from the surface up. To interconnect the layers, one can make use of covalent bivalent chemical linkers, but biospecific interactions and electrostatic interactions are more commonly employed.

7.7.1 LbL Assembly of VNPs Using Biospecific Interactions

Multi-layered thin films of CCMV and CPMV have been constructed using the biotin–streptavidin system (Steinmetz *et al.*, 2008a, 2006; Suci *et al.*, 2006). To achieve this, biotinylated VNPs are immobilized on a solid support; biotinylated VNPs can be bound on a streptavidin-coated substrate or Cys-added biotinylated mutants bound on a gold surface. Next, strepatividin is added. Streptavidin is a tetrameric protein and hence has four binding sites for biotin. The multivalency of the VNPs and streptavidin allows interconnection of the VNP layers. A second biotinylated VNP can then be deposited, followed by a further streptavidin layer, and so on. This process can be repeated until the desired number of VNP layers is deposited.

In a proof-of-concept study, CPMV particles were covalently labeled with two different ligands: biotin moieties were displayed allowing self-assembly via biospecific interaction with streptavidin, and displayed fluorescent labels served as imaging molecules (Steinmetz *et al.*, 2006). Attachment of the different functionalities was achieved using the design principles described in Chapter 4. Different color-labeled sets of biotinylated CPMV particles were added sequentially and an array was built from the surface up (Steinmetz *et al.*, 2006). Fluorescence microscopic analysis allowed differentiation between a mixed monolayer *versus* a bilayer structure (Fig. 7.19) (Steinmetz *et al.*, 2006).

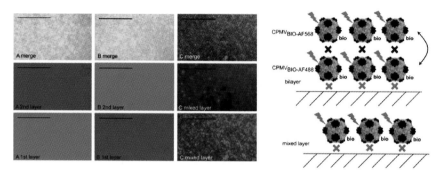

Figure 7.19 Bilayers and a mixed monolayer of biotinylated (bio) and fluorescent-labeled CPMV particles on Au slides imaged via fluorescence microscopy (left), and diagrammatic representation of layer structures (right). The green and red flags show the AlexaFluor dyes AF488 and AF568, respectively. The black cross depicts streptavidin (SAv); the gray cross shows a thiol-modified SAv. The scale bar is 10 μm. (a) Bilayer of CPMV-BIO-AF488 and CPMV-BIO-AF568, CPMV-BIO-AF488 in the first and CPMV-BIO-AF568 in the second layer; merge shows the overlaid images from the first and second layers. (b) Bilayer of CPMV-BIO-AF568 and CPMV-BIO-AF488. (c) Mixed monolayer of CPMV-BIO-AF488 and CPMV-BIO-AF568. Reproduced with permission from Steinmetz, N. F., Calder, G., Lomonossoff, G. P., and Evans, D. J. (2006) Plant viral capsids as nanobuilding blocks: construction of arrays on solid supports, *Langmuir*, **22**(24), 10032–10037.

In a follow-up study, the mechanical properties of such arrays were investigated (Steinmetz *et al.*, 2008a). Multi-layered build-up of various CPMV building blocks modified with varying densities (30 labels versus 200 labels) of biotin molecules attached via longer (3.50 nm) and shorter linkers (2.24 nm) was investigated. Arrays were fabricated by alternating deposition of streptavidin and biotinylated CPMV particles. Biophysical measurements showed that the mode of multi-layer assembly was different for each array fabricated. The general trend was that a more regular and densely packed, strongly interlinked array was assembled when CPMV particles displaying a high number of biotin labels attached via the longer linker were used. In stark contrast were the mechanical properties of arrays formed with CPMV particles displaying a low number of biotins attached via the shorter linker. The formation of a viscoelastic array with less dense particle packaging was implied (Steinmetz *et al.*, 2008a). The observation that the mechanical properties of the CPMV arrays can be tightly tuned by adjusting spacer length and density is interesting and important. The judicious use of appropriate linkers will allow the design of thin films with the desired needs and properties, for example, rigid and densely packed versus viscoelastic and less densely packed.

The biotin–streptavidin system in combination with the hexa-His–Ni–NTA system has also been used to construct hybrid arrays consisting of VNPs and inorganic materials. Quantum dots (QDs) have been specifically captured onto an immobilized CPMV layer (Medintz *et al.*, 2005). The CPMV layer offers a large surface area compared with a flat surface. In addition, the CPMV layer offers symmetrically arranged anchoring groups. It was estimated that when a close-packed layer of CPMV was used to capture QDs, a 40% increase in QD density could be achieved. An enhanced fluorescence signal was indeed observed comparing quantum dots captured on a CPMV surface versus a flat surface (Medintz *et al.*, 2005).

The utility of such a hybrid CPMV system for data storage has been demonstrated (Portney *et al.*, 2007). CPMV particles decorated with QDs were immobilized and embedded in a polyvinyl alcohol (PVA) matrix. The PVA matrix containing the hybrid VNPs was sandwiched between two electrodes and an electric field was applied. The current–voltage curves showed several multi-stable switching states. This implies that these materials could indeed find applications in non-volatile memory devices (non-volatile describes a storage device that stores information even when it is not powered) (Portney *et al.*, 2007).

These techniques have been extended and iron oxide nanoparticles have also been captured on CPMV monolayers by making use of covalent chemical linkages. Potential applications of CPMV–iron oxide nanoparticle hybrids lie in the field of bioimaging, such as MRI (see Chapter 8) (Martinez-Morales *et al.*, 2008) .

7.7.2 LbL Assembly Using Electrostatic Interactions

Multi-layered arrays consisting of polyelectrolytes and VNPs can be fabricated via electrostatic interactions. Interestingly, the behavior of icosahedrons versus rods in polyelectrolyte multilayers was found to be different. Multi-layered arrays of alternating charged polyelectrolytes and charged icosahedral VNPs, such as CCMV, CPMV, or *Carnation mottle virus* (CarMV),[2] have been successfully self-assembled using the LbL approach (Lin, 2008;

---

[2] CarMV is a plant virus in the Carmovirus genus. The particles have icosahedral $T = 3$ symmetry and a diameter of about 28 nm. CarMV is a single-stranded RNA virus. It naturally occurs in plants from the family *Caryophyllaceae* and is widespread and common in carnation [from Description of Plant Viruses (DPV): www.dpvweb.net].

Lvov *et al.*, 1994; Steinmetz *et al.*, 2008b; Suci *et al.*, 2005, 2006). Figure 7.20 depicts a multi-layered structure of CPMV particles and the polymer pair linear poly(ethyleneimine) (PEI) and poly(acrylic acid) (PAA). CPMV particles have a negative surface charge (under the conditions employed) and bind stably to the positively charged PEI. An alternating structure of polyelectrolyte with incorporated CPMV nanoparticles was self-assembled (Steinmetz *et al.*, 2008b).

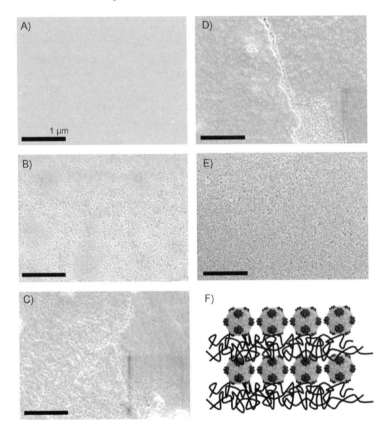

Figure 7.20 Scanning electron micrographs showing the sequential build-up of polyelectrolytes and cowpea mosaic virus (CPMV) particles. (a) Precursor thin film consisting of 2.5 bilayers of the polyions linear poly(ethyleneimine) (PEI) and poly(acrylic acid) (PAA); (b) (PEI–PAA)$_{2.5}$ CPMV; (c) (PEI–PAA)$_{2.5}$ CPMV (PEIPAA); (d) (PEI–PAA)$_{2.5}$ CPMV (PEI–PAA)$_3$; (e) (PEI–PAA)$_{2.5}$ CPMV (PEI–PAA)$_3$ CPMV; (f) schematic representation of the architecture. All images were viewed at an acceleration voltage of 5 kV with a Zeiss Supra 55 VP FEG SEM; scale bar: 1 µm. Reproduced with permission from Steinmetz, N. F., Findlay, K. C., Noel, T. R., Parker, R., Lomonossoff, G. P., and Evans, D. J. (2008) Layer-by-layer assembly of viral nanoparticles and polyelectrolytes: the film architecture is different for spheres versus rods, *ChemBioChem*, **9**(10), 1662–1670.

Whereas icosahedral VNPs are incorporated into polyelectrolyte multi-layered arrays, rod-shaped VNPs are not stably incorporated into an alternating array, but assemble on top of the array. Rod-shaped particles such as M13 and TMV were found to "float" on top of the polyelectrolyte multi-layered array; they are excluded from the array and assemble in a highly organized structure at the interface (Steinmetz *et al.*, 2008b; Yoo *et al.*, 2006).

The "floating" phenomenon can be explained by inter-diffusion of the polyelectrolytes (see below). The first study describing this phenomenon was with M13 and the polyelectrolytes PEI and PAA. The mechanism is outlined in Fig. 7.21 (Yoo *et al.*, 2006).

Step 1: A supporting film of polyelectrolytes is assembled using the LbL technique; the terminating layer is the positively charged PEI. M13 is then introduced to the system. M13 particles are negatively charged under the conditions used, and thus bind to the positively charged terminating PEI layer. M13 particles are randomly distributed and highly disordered.

Step 2: A further PEI layer is added. Some M13 particles dislocate and appear on top of the structure.

Step 3: The negatively charged PAA is added. The VNPs are covered under the PAA layer, also described as *blanketing.*

Step 4: PEI is added. This again results in the dislocation of M13 and the particles appear on top of the assembly.

Step 5: The cycles can be repeated until a desired number of PEI–PAA layers are deposited. With every PEI layer added to the system, M13 particles dislocate and appear on top of the array. The diffusion process also leads to an alignment of the particles at the interface. Eventually, a highly organized 2D array of M13 particles is fabricated (Fig. 7.21).

The "floating" mechanism can be explained by inter-diffusion of the polyelectrolytes (Yoo *et al.*, 2006). PEI plays a major role in the inter-diffusion process. It diffuses under the M13 layer into the multi-layered matrix and outwards to the virus-covered surface. The PEI inter-diffusion results in dislocation of the M13 particles and their alignment on the surface. The ordering of the VNPs into a 2D array is spontaneous and induced by repulsive electrostatic interactions.

The "floating" mechanism has so far only been observed with rod-shaped VNPs (Steinmetz *et al.*, 2008b; Yoo *et al.*, 2006). In physics, buoyancy, the upward force of an object in a fluid, is dependent on the density of the fluid and the density and the surface area of the object itself. The densities of rods and icosahedrons are not significantly different. In stark contrast are the surface areas of each particle type. The floating TMV and M13 rods have a surface area of 22,200 and 18,000 nm², respectively, whereas the typical icosahedron CPMV has a surface area of only 2800 nm². This difference of almost an order of magnitude would be expected to have a significant impact on the physical properties of the particles both in solution and when embedded in flexible and tunable polyelectrolyte multilayers. The striking structural differences may explain why rods float and icosahedrons do not.

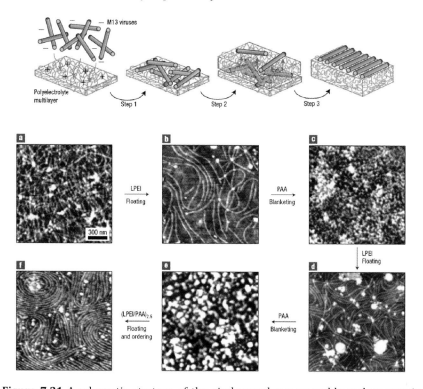

Figure 7.21 A schematic strategy of the viral monolayer assembly and an atomic force microscopic (AFM) demonstration of the disordered state and the ordered monolayer state of the M13 VNP. Top panel: an experimental procedure for monolayer assembly of M13 virus on the polyelectrolyte multilayer of PEI/PAA. Bottom panel: a series of images to investigate the inter-diffusion process in the PEI/PAA system. The 1.5 × 1.5 µm² height-mode AFM images are presented (Z-range, height scale, 20 nm). All of the species are deposited at pH 5.0. (a) Initial disorderly adsorbed viruses on (PEI/PAA)3.5. (b–e) Alternating depositions of PEI and PAA onto the prepared VNP layer of (a). (f) After further deposition of (PEI/PAA)4.5 onto the surface of (e). A highly ordered and closely packed monolayer of M13 is obtained on the surface. The scale bar in (a) refers to all parts. Reproduced with permission from Yoo, P. J., Nam, K. T., Qi, J. F., Lee, S. K., Park, J., Belcher, A. M., and Hammond, P. T. (2006) Spontaneous assembly of viruses on multilayered polymer surfaces, *Nat. Mater.*, **5**(3), 234–240.

The phenomenon of spontaneous self-assembly of rods at the polyelectrolyte interface has resulted in useful techniques. Battery electrodes were fabricated using assemblies of metalized M13 particles organized on polyelectrolyte films (as discussed in Chapter 6) (Lee *et al.*, 2009; Nam *et al.*, 2006, 2008). This assembly can also be combined with molding techniques and 2D patterns of polymeric M13 films can be fabricated (Yoo *et al.*, 2008).

Multi-layered arrays of VNPs fabricated via electrostatic interactions offer high tunability and flexibility. The interactions between the VNPs and charged polyelectrolyes can be modulated by ionic strength and pH; the array can potentially be disrupted under certain conditions, thus releasing the VNPs. This may be useful for the development of delivery systems for biomedical applications (discussed in Chapter 8).

7.8 VNP ARRAY FORMATION IN SOLUTION

Arrays of VNPs can be formed in solution. However, because of multivalency of VNPs, array formation and growth are difficult to control in solution and often lead to the formation of aggregates. Nevertheless, few examples demonstrate VNP network formulation in solution: Two sets of CPMV particles, each decorated with complementary oligonucleotides, were used to promote self-assembly in solution via nucleic acid hybridization. The system is reversible upon addition of a competing oligonucleotide (Strable *et al.*, 2004). Arrays of VNPs in solution can also be formed by mixing biotinylated VNPs with streptavidin (Wang *et al.*, 2002).

Hybrid networks of VNPs and inorganic materials have also been fabricated in solution. Aggregates of CPMV and QDs as well as *Flock house virus* and carbon nanotubes have been generated using covalent bioconjugation strategies (Portney *et al.*, 2005). Non-covalent hybrid networks have been created using genetically engineered M13 particles. Here, a phage construct expressing a gold-binding peptide as well as tissue-specific targeting peptides was mixed with gold. The interaction with gold led to the formation of a hybrid network that has been utilized for cell targeting and imaging applications and will be discussed in Chapter 8 (Souza *et al.*, 2006).

7.9 ARRAYS OF ENVELOPED VNPs

The infection process of enveloped VNPs involves fusion of the viral membrane with cellular membranes such as the plasma membrane or the endosomal membrane. Fusion with endosomal membranes is often driven by low pH and results in release of the genomic information into the cytoplasm. This mechanism has been utilized to trigger fusion of enveloped VNPs with lipid membranes on solid supports and beads, where surface is coated with a lipid bilayer followed by low-pH-catalyzed fusion. Systems with stably immobilized *Rubella* VLPs and *Influenza* A/PR8 viruses[3] have been developed (Fischlechner *et al.*, 2006, 2007; Toellner *et al.*, 2006).

Such virus-coated beads have been used as diagnostic tools for the detection of virus-specific antibodies. To highlight an example, *Rubella* VLPs or *Influenza* A/PR8 viruses were stably embedded onto beads coated with lipid bilayers. First, the beads were coated with the polyelectrolytes poly(allylamine hydrochloride) (PAH) and poly(styrene sulfonate) (PSS). Next, a lipid mixture of phosphatidylserine (PS) and phosphatidylcholine (PC) was bound. When *Rubella* VLPs or *Influenza* A/PR8 viruses were added to these beads at low pH, the particles fused with the lipid bilayer and, as a result, the VLPs were stably and firmly incorporated in the supported membrane (Fischlechner *et al.*, 2007). To demonstrate the feasibility of such VLP-coated beads in a diagnostic assay to detect virus-specific antibodies, two different sets of beads were immobilized on a surface to create a "sensor-on-a-chip" device. Microcontact printing was used to assemble *Rubella* VLP-coated beads and *Influenza* A/PR8 virus-coated beads on a chip (Fig. 7.22) (Fischlechner *et al.*, 2007). *Rubella* VLPs served as a negative control and were labeled in green by fluorescent dye incorporation; *Influenza* A/PR8 virus-coated beads were not color-coded. The device was then exposed to polyclonal sera against *Influenza* A/PR8 virus. Detection was carried out using phycoerythrin-labeled secondary antibodies (red) (Fig. 7.22) (Fischlechner *et al.*, 2007). Immunofluorescent imaging showed high specificity, as indicated by the presence of very few yellow spots (yellow spots result from an overlay of red and green fluorescence and are false positive). The beads were further tested using flow cytometry; the data were consistent, confirming high selectivity (Fischlechner *et al.*, 2007). The methods established could be applied to other pathogens and may lead to the development of a range of devices for use in diagnostics.

---

[3]*Influenza* viruses are the infectious agents causing the disease influenza, commonly referred to as the flu. *Influenza* viruses belong to the family *Orthomyxoviridae*, and affects birds and mammals. The strain utilized is *Influenza* A/PR8 (H1N1), which has been isolated in Puerto Rico. The virions have a complex structure and are enveloped. Virions are spherical to pleomorphic and around 80–120 nm in diameter and 200–300 nm in length. *Influenza* A/PR8 (H1N1) has a segmented genome consisting of eight segments of linear, negative-sense, single-stranded RNA [from the Database of the International Committee on taxonomy on Viruses (ICTV; http://www.ncbi.nlm.nih.gov/ICTVdb)].

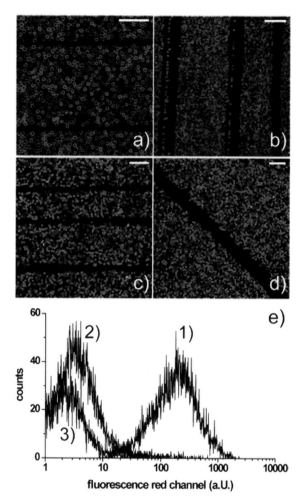

Figure 7.22 2D arrangement of composite LbL-lipid-virus colloids on slides patterned with PAH. (a, b) Lipid-coated colloids arranged on a PAH-printed glass slide. (c, d) Immunofluorescent assay on colloids fused with RV VLPs (green) as negative control and Influenza A/PR8-coated colloids. The red signal indicates specific binding of the primary antibodies as well as the secondary antibody on colloids fused with Influenza A/PR8. Scale bars: 25 µm. (e) Immunofluorescence assay on colloids using flow cytometry. (e, assay 1) Colloids fused with Influenza A/ PR8 [geometric mean fluorescence (gmf), 163]. Controls: colloids fused with RLPs (e, assay 2, gmf, 3.1), and lipid-coated colloids (e, assay 3, gmf, 2.5). Self-fluorescence of lipid-coated colloids was 2.3 (data not shown). Reproduced with permission from Fischlechner, M., Reibetanz, U., Zaulig, M., Enderlein, D., Romanova, J., Leporatti, S., Moya, S., and Donath, E. (2007) Fusion of enveloped virus nanoparticles with polyelectrolyte-supported lipid membranes for the design of bio/non-bio interfaces, *Nano Lett.*, **7**(11), 3540–3546.

7.10 CONCLUSIONS

A variety of techniques have been developed allowing the fabrication of VNP structures including 1D bundles and wires, 2D films, and multi-layered arrays leading toward 3D structures. VNPs can be immobilized on nearly any material and complex structures can be assembled. Once assembled, the VNPs can serve as platforms for further functionalization. Alternatively, VNPs displaying functional moieties can be immobilized and assembled into devices. The potential applications are open-ended and range from data storage and battery electrodes to bioimaging and diagnostics.

References

Chatterji, A., Ochoa, W. F., Ueno, T., Lin, T., and Johnson, J. E. (2005) A virus-based nanoblock with tunable electrostatic properties, *Nano Lett.*, **5**(4), 597–602.

Cheung, C. L., Camarero, J. A., Woods, B. W., Lin, T., Johnson, J. E., and De Yoreo, J. J. (2003) Fabrication of assembled virus nanostructures on templates of chemoselective linkers formed by scanning probe nanolithography, *J. Am. Chem. Soc.*, **125**(23), 6848–6849.

Cheung, C. L., Chung, S. W., Chatterji, A., Lin, T., Johnson, J. E., Hok, S., Perkins, J., and De Yoreo, J. J. (2006) Physical controls on directed virus assembly at nanoscale chemical templates, *J. Am. Chem. Soc.*, **128**(33), 10801–10807.

de la Escosura, A., Verwegen, M., Sikkema, F. D., Comellas-Aragones, M., Kirilyuk, A., Rasing, T., Nolte, R. J., and Cornelissen, J. J. (2008) Viral capsids as templates for the production of monodisperse Prussian blue nanoparticles, *Chem. Commun. (Camb)*, (13), 1542–1544.

Fang, J., Soto, C. M., Lin, T., Johnson, J. E., and Ratna, B. R. (2002) Complex pattern formation of Cowpea mosaic virus nanoparticles, *Langmuir*, **18**, 308–310.

Fischlechner, M., Reibetanz, U., Zaulig, M., Enderlein, D., Romanova, J., Leporatti, S., Moya, S., and Donath, E. (2007) Fusion of enveloped virus nanoparticles with polyelectrolyte-supported lipid membranes for the design of bio/nonbio interfaces, *Nano Lett.*, **7**(11), 3540–3546.

Fischlechner, M., Toellner, L., Messner, P., Grabherr, R., and Donath, E. (2006) Virus-engineered colloidal particles – a surface display system, *Angew. Chem. Int. Ed.*, **45**(5), 784–789.

Gentili, M., Giovannella, C., and Selci, S. (2008) *Nanolithography: A Borderland Between STM, EB, IB, and X-ray Lithographies*, Kluwer Academic Publishers, Dordrecht, the Netherlands.

He, J., Niu, Z., Tangirala, R., Wang, J. Y., Wei, X., Kaur, G., Wang, Q., Jutz, G., Boker, A., Lee, B., Pingali, S. V., Thiyagarajan, P., Emrick, T., and Russell, T. P. (2009) Self-assembly of tobacco mosaic virus at oil/water interfaces, *Langmuir*, **25**(9), 4979–4987.

Huang, Y., Chiang, C. Y., Lee, S. K., Gao, Y., Hu, E. L., De Yoreo, J., and Belcher, A. M. (2005) Programmable assembly of nanoarchitectures using genetically engineered viruses, *Nano Lett.*, **5**(7), 1429–1434.

Klem, M. T., Willits, D., Young, M., and Douglas, T. (2003) 2-D array formation of genetically engineered viral cages on Au surfaces and imaging by atomic force microscopy, *J. Am. Chem. Soc.*, **125**(36), 10806–10807.

Knez, M., Bittner, A. M., Boes, F., Wege, C., Jeske, H., Maiß, E., and Kern, K. (2003) Biotemplate synthesis of 3-nm nickel and cobalt nanowires, *Nano Lett.*, **3**(8), 1079–1082.

Knez, M., Sumser, M., Bittner, A. M., Wege, C., Jeske, H., Hoffmann, D. M., Kuhnke, K., and Kern, K. (2004) Binding the tobacco mosaic virus to inorganic surfaces, *Langmuir*, **20**(2), 441–447.

Kuncicky, D. M., Naik, R. R., and Velev, O. D. (2006) Rapid deposition and long-range alignment of nanocoatings and arrays of electrically conductive wires from tobacco mosaic virus, *Small*, **2**(12), 1462–1466.

Lee, S. K., Yun, D. S., and Belcher, A. M. (2006) Cobalt ion mediated self-assembly of genetically engineered bacteriophage for biomimetic Co–Pt hybrid material, *Biomacromolecules*, **7**(1), 14–17.

Lee, S. W., and Belcher, A. M. (2004) Virus-based fabrication of micro- and nanofibers using electrospinning, *Nano Lett.*, **4**(3), 387–390.

Lee, S. W., Lee, S. K., and Belcher, A. M. (2003a) Virus-based alignment of inorganic, organic, and biological nanosized materials, *Adv. Mater.*, **15**, 689–692.

Lee, S. W., Mao, C., Flynn, C. E., and Belcher, A. M. (2002) Ordering of quantum dots using genetically engineered viruses, *Science*, **296**(5569), 892–895.

Lee, S. W., Woods, B. W., and Belcher, A. M. (2003b) Chiral smectic C structures of virus-based films, *Langmuir*, **19**, 1592–1598.

Lee, Y. J., Yi, H., Kim, W. J., Kang, K., Yun, D. S., Strano, M. S., Ceder, G., and Belcher, A. M. (2009) Fabricating genetically engineered high-power lithium-ion batteries using multiple virus genes, *Science*, **324**(5930), 1051–1055.

Lvov, Y., Haas, H., Decher, G., and Möhwald, H. (1994) Successive deposition of alternate layers of polyelectrlytes and a charged virus, *Langmuir*, **10**, 4232–4236.

Martinez-Morales, A. A., Portney, N. G., Zhang, Y., Destito, G., Budak, G., Ozbay, E., Manchester, M., Ozkan, C. S., and Ozkan, M. (2008) Synthesis and characterization of iron oxide derivatized mutant Cowpea mosaic virus hybrid nanoparticles, *Adv. Mater.*, **20**, 1–5.

Medintz, I. L., Sapsford, K. E., Konnert, J. H., Chatterji, A., Lin, T., Johnson, J. E., and Mattoussi, H. (2005) Decoration of discretely immobilized cowpea mosaic virus with luminescent quantum dots, *Langmuir*, **21**(12), 5501–5510.

Nam, K. T., Kim, D. W., Yoo, P. J., Chiang, C. Y., Meethong, N., Hammond, P. T., Chiang, Y. M., and Belcher, A. M. (2006) Virus-enabled synthesis and assembly of nanowires for lithium ion battery electrodes, *Science*, **312**(5775), 885–888.

Nam, K. T., Peelle, B. R., Lee, S. W., and Belcher, A. M. (2004) Genetically driven assembly of nanorings based on the M13 virus, *Nano Lett.*, **4**(1), 23–27.

Nam, K. T., Wartena, R., Yoo, P. J., Liau, F. W., Lee, Y. J., Chiang, Y. M., Hammond, P. T., and Belcher, A. M. (2008) Stamped microbattery electrodes based on self-assembled M13 viruses, *Proc. Natl Acad. Sci. USA*, **105**(45), 17227–17231.

Niu, Z., Bruckman, M. A., Li, S., Lee, L. A., Lee, B., Pingali, S. V., Thiyagarajan, P., and Wang, Q. (2007a) Assembly of tobacco mosaic virus into fibrous and macroscopic bundled arrays mediated by surface aniline polymerization, *Langmuir*, **23**(12), 6719–6724.

Niu, Z., Liu, J., Lee, L. A., Bruckman, M. A., Zhao, D., Koley, G., and Wang, Q. (2007b) Biological templated synthesis of water-soluble conductive polymeric nanowires, *Nano Lett.*, **7**(12), 3729–3733.

Portney, N. G., Singh, K., Chaudhary, S., Destito, G., Schneemann, A., Manchester, M., and Ozkan, M. (2005) Organic and inorganic nanoparticle hybrids, *Langmuir*, **21**(6), 2098–2103.

Portney, N. G., Tseng, R. J., Destito, G., Strable, E., Yang, Y., Manchester, M., Finn, M. G., and Ozkan, M. (2007) Microscale memory characteristics of virus-quantum dot hybrids, *Appl. Phys. Lett.*, **90**, 214104.

Royston, E., Ghosh, A., Kofinas, P., Harris, M. T., and Culver, J. N. (2008) Self-assembly of virus-structured high surface area nanomaterials and their application as battery electrodes, *Langmuir*, **24**(3), 906–912.

Royston, E. S., Brown, A. D., Harris, M. T., and Culver, J. N. (2009) Preparation of silica stabilized Tobacco mosaic virus templates for the production of metal and layered nanoparticles, *J. Colloid Interface Sci.*, **332**(2):, 402–407.

Russell, J. T., Lin, Y., Boker, A., Su, L., Carl, P., Zettl, H., He, J., Sill, K., Tangirala, R., Emrick, T., Littrell, K., Thiyagarajan, P., Cookson, D., Fery, A., Wang, Q., and Russell, T. P. (2005) Self-assembly and cross-linking of bionanoparticles at liquid–liquid interfaces, *Angew. Chem. Int. Ed. Engl.*, **44**(16), 2420–2426.

Shenton, W., Douglas, T., Young, M., Stubbs, G., and Mann, S. (1999) Inorganic–organic nanotube composites from template mineralization of Tobacco mosaic virus, *Adv. Mater.*, **11**, 253–265.

Smith, J. C., Lee, K., Wang, Q., Finn, M. G., Johnson, J. E., Mrksich, M., and Mirkin, C. A. (2003) Nanopatterning the chemospecific immobilization of Cowpea mosaic virus capsid, *Nano Lett.*, **3**, 883–886.

Souza, G. R., Christianson, D. R., Staquicini, F. I., Ozawa, M. G., Snyder, E. Y., Sidman, R. L., Miller, J. H., Arap, W., and Pasqualini, R. (2006) Networks of gold nanoparticles and bacteriophage as biological sensors and cell-targeting agents, *Proc. Natl Acad. Sci. USA*, **103**, 1215–1220.

Steinmetz, N. F., Bock, E., Richter, R. P., Spatz, J. P., Lomonossoff, G. P., and Evans, D. J. (2008a) Assembly of multilayer arrays of viral nanoparticles via biospecific recognition: a quartz crystal microbalance with dissipation monitoring study, *Biomacromolecules*, **9**(2), 456–462.

Steinmetz, N. F., Calder, G., Lomonossoff, G. P., and Evans, D. J. (2006) Plant viral capsids as nanobuilding blocks: construction of arrays on solid supports, *Langmuir*, **22**(24), 10032–10037.

Steinmetz, N. F., Findlay, K. C., Noel, T. R., Parker, R., Lomonossoff, G. P., and Evans, D. J. (2008b) Layer-by-layer assembly of viral nanoparticles and polyelectrolytes: the film architecture is different for spheres versus rods, *ChemBioChem*, **9**(10), 1662–1670.

Strable, E., Johnson, J. E., and Finn, M. G. (2004) Natural nanochemical building blocks: icosahedral virus particles organized by attached oligonucleotides, *Nano Lett.*, **4**, 1385–1389.

Suci, P. A., Klem, M. T., Arce, F. T., Douglas, T., and Young, M. (2006) Assembly of multilayer films incorporating a viral protein cage architecture, *Langmuir*, **22**(21), 8891–8896.

Suci, P. A., Klem, M. T., Douglas, T., and Young, M. (2005) Influence of electrostatic interactions on the surface adsorption of a viral protein cage, *Langmuir*, **21**(19), 8686–8693.

Suh, K. Y., Khademhosseini, A., Jon, S., and Langer, R. (2006) Direct confinement of individual viruses within polyethylene glycol (PEG) nanowells, *Nano Lett.*, **6**(6), 1196–1201.

Sun, J., DuFort, C., Daniel, M. C., Murali, A., Chen, C., Gopinath, K., Stein, B., De, M., Rotello, V. M., Holzenburg, A., Kao, C. C., and Dragnea, B. (2007) Core-controlled polymorphism in virus-like particles, *Proc. Natl Acad. Sci. USA*, **104**(4), 1354–1359.

Tan, W. S., Lewis, C. L., Horelik, N. E., Pregibon, D. C., Doyle, P. S., and Yi, H. (2008) Hierarchical assembly of viral nanotemplates with encoded microparticles via nucleic acid hybridization, *Langmuir*, **24**(21), 12483–12488.

Toellner, L., Fischlechner, M., Ferko, B., Grabherr, R. M., and Donath, E. (2006) Virus-coated layer-by-layer colloids as a multiplex suspension array for the detection and quantification of virus-specific antibodies, *Clin. Chem.*, **52**(8), 1575–1583.

Udit, A. K., Brown, S., Baksh, M. M., and Finn, M. G. (2008) Immobilization of bacteriophage Qbeta on metal-derivatized surfaces via polyvalent display of hexahistidine tags, *J. Inorg. Biochem.*, **102**(12), 2142–2146.

Vega, R. A., Maspoch, D., Salaita, K., and Mirkin, C. A. (2005) Nanoarrays of single virus particles, *Angew. Chem. Int. Ed. Engl.*, **44**(37), 6013–6015.

Wang, Q., Kaltgrad, E., Lin, T., Johnson, J. E., and Finn, M. G. (2002) Natural supramolecular building blocks: wild-type cowpea mosaic virus, *Chem. Biol.*, **9**(7), 805–811.

Wang, X., Yan, Y., Yost, M. J., Fann, S. A., Dong, S., and Li, X. (2007) Nanomechanical characterization of micro/nanofiber reinforced type I collagens, *J. Biomed. Mater. Res. A*, **83**(1), 130–135.

Wargacki, S. P., Pate, B., and Vaia, R. A. (2008) Fabrication of 2D ordered films of tobacco mosaic virus (TMV): processing morphology correlations for convective assembly, *Langmuir*, **24**(10), 5439–5444.

Yi, H., Nisar, S., Lee, S. Y., Powers, M. A., Bentley, W. E., Payne, G. F., Ghodssi, R., Rubloff, G. W., Harris, M. T., and Culver, J. N. (2005) Patterned assembly of genetically modified viral nanotemplates via nucleic acid hybridization, *Nano Lett.*, **5**(10), 1931–1936.

Yi, H., Rubloff, G. W., and Culver, J. N. (2007) TMV microarrays: hybridization-based assembly of DNA-programmed viral nanotemplates, *Langmuir*, **23**(5), 2663–2667.

Yoo, P. J., Nam, K. T., Belcher, A. M., and Hammond, P. T. (2008) Solvent-assisted patterning of polyelectrolyte multilayers and selective deposition of virus assemblies, *Nano Lett.*, **8**(4), 1081–1089.

Yoo, P. J., Nam, K. T., Qi, J. F., Lee, S. K., Park, J., Belcher, A. M., and Hammond, P. T. (2006) Spontaneous assembly of viruses on multilayered polymer surfaces, *Nat. Mater.*, **5**(3), 234–240.

Chapter 8

VNPs AS TOOLS FOR NANOMEDICINE

The development of nanomaterials for medical applications is termed *nanomedicine.* Nanomaterials have been employed for applications such as the fabrication of novel sensors for medical diagnostics, or the development of nanodevices for tissue-specific imaging and therapy (reviewed in Jain, 2008; Maham *et al..*, 2009; Sandhiya *et al.*, 2009). Imaging modalities and drugs are typically small chemicals that lack tissue specificity. These molecules therefore show broad tissue distribution and accumulate in diseased as well as in healthy tissues, thus reducing the sensitivity or leading to undesired side effects. The doses of imaging or therapeutic molecules that can be used are therefore limited.

The ability to specifically target imaging and/or therapeutic compounds to disease sites is regarded an important goal in nanomedicine that would increase imaging sensitivity, increase efficacy of chemotherapeutics, and reduce side effects. Nanotechnology has opened the door for the development of "smart" targeted devices by interlinking targeting ligands with imaging and/or therapeutic molecules, and holds great promise for the development of improved therapies, especially for treatment of cancer or vascular disease.

Diverse classes of novel nanomaterials have recently been developed, including nanocrystals and quantum dots (QDs), dendrimers, polymer vesicles and beads, liposomes, and protein-based nanostructures such as protein cages and viral nanoparticles (VNPs) (reviewed in Bawarski *et al.*, 2008; Majoros *et al.*, 2008; Manchester & Singh, 2006; Soussan *et al.*, 2009; Xing & Rao, 2008). Each of these systems has advantages and disadvantages regarding biocompatibility, bioavailability and pharmacokinetics, toxicity, immunogenicity, and specificity for the target tissue. VNPs have many appealing features and offer various advantages compared to synthetic nanomaterials. The main benefit is that they are naturally biocompatible and biodegradable. Biocompatibility is particularly important for medical

Viral Nanoparticles: *Tools for Materials Science and Biomedicine*
By Nicole F. Steinmetz and Marianne Manchester
Copyright © 2011 by Pan Stanford Publishing Pte. Ltd.
www.panstanford.com

applications. In particular, non-human pathogens such as VNPs from plant viruses or bacteriophages are less likely to interact with human receptors or trigger signal transduction events that may lead to downstream side effects. In general, replication or gene expression of plant VNPs or bacteriophages is not supported in mammalian systems.

The use of VNPs as platforms for the design of novel formulations for use in medicine has attracted many researchers, and the field is rapidly evolving. This chapter summarizes recent advances in the use and development of VNPs for potential biomedical applications. The first approaches toward utilization of VNPs and virus-like particles (VLPs) in medicine focused on vaccines (see Section 8.2). More recent developments utilized labeled VNPs as imaging sensors (see Section 8.3). With regard to targeted therapies, a range of targeted drug-carrying VNPs have been developed and tested (see Sections 8.4 and 8.5). Last but not least, VNPs and VLPs are exploited as vehicles for therapeutic gene delivery (see Section 8.6). Although virus-mediated gene delivery is a very large field and beyond the scope of this chapter, we will focus on some recent discoveries in this area.

8.1 TOXICITY, BIODISTRIBUTION, PHARMACOKINETICS, AND IMMUNOGENICITY OF VNPs

When developing novel materials for potential applications in medicine it is essential to understand the *in vivo* properties of the material, including pharmacokinetics, clearance and elimination, tissue biodistribution, immunogenicity, and toxicity. Factors such as immunogenicity and potential toxic effects must be elucidated in detail. Plant VNPs, VLPs, and phages can be considered as safe from a human health perspective. Nevertheless, as of today, there are only a few studies describing the characterization of VNP platforms *in vivo*, as summarized below.

8.1.1 Dosage and Potential Toxicity

The platforms *Adenovirus* (Ad), adeno-associated virus (AAV), and *Lentivirus* have long been investigated as vectors for therapeutic gene delivery. One major problem with these human pathogens is that, even when using replication-deficient versions, toxic effects are induced. These platforms cause acute induction of pro-inflammatory responses, and a dosage of 2×10^{11} Ads/kg body weight has been found to induce substantial liver injury (Ben-Gary *et al.*, 2002; Christ *et al.*, 2000; Cotter & Muruve, 2005; Engler *et al.*, 2004; Higginbotham *et al.*, 2002; Reid *et al.*, 2002).

VNPs derived from plant viruses and phages are less likely to interact specifically with the mammalian system and therefore less likely to cause potential toxic side effects. For *Cowpea mosaic virus* (CPMV), it was shown that, even up to dosages of 10^{16} CPMV particles per kilogram body weight, no apparent toxic side effects were observed (Singh *et al.*, 2007). Animals injected intravenously with 1, 10, and 100 mg/kg CPMV did not show any apparent clinical signs, or evidence of toxicity (Singh *et al.*, 2007). These findings were in good agreement with studies using CCMV, where *Cowpea chlorotic mottle virus* (CCMV)-injected animals did not show any clinical symptoms and did not have any histological pathology (Kaiser *et al.*, 2007).

VNPs are immunogenic materials, and it was suggested that they are eliminated via B-lymphocytes (these cells are part of the humoral immune system, as opposed to the cell-mediated immune system). In the case of some VNPs (CPMV, T7, and Qβ), it has been shown that administration leads to a transient increase in the B-cell numbers; however, animals recover from this response within a few days, suggesting clearance of the VNPs (Gatto *et al.*, 2004; Singh *et al.*, 2007; Srivastava *et al.*, 2004). It was indicated that CPMV particles persist for at least 3 days (Rae *et al.*, 2005). In contrast, CCMV particles are nearly completely eliminated within 24 h after administration (Kaiser *et al.*, 2007). Data from longer-term studies have not been reported yet.

8.1.2 Plasma Clearance, Bioavailability, and Biodistribution

Pharmacokinetics and tissue biodistribution are important factors to evaluate *in vivo*, and characteristics such as particle size, surface charge, and alteration play a crucial role. It is important to find the right balance between tissue penetration and systemic clearance. Longer circulation times allow greater specificity and better accumulation in the target tissue; however, more rapid clearance might be desired to reduce toxic side effects or background for imaging (Thurber *et al.*, 2008).

Both CPMV and CCMV have broad biodistribution and were detected in a wide variety of tissues throughout the body with no apparent toxic effects (Kaiser *et al.*, 2007; Rae *et al.*, 2005). CPMV particles mostly accumulated in liver and spleen (Singh *et al.*, 2007). This distribution is similar to observations made using Qβ and M13 (Molenaar *et al.*, 2002; Prasuhn *et al.*, 2008). A different pattern was observed with CCMV; CCMV particles were mostly found in the thyroid as well as in liver, spleen, bladder, and salivary gland (Kaiser *et al.*, 2007). Uptake and accumulation of VNPs in organs with filtration function, such as liver and spleen, were expected. These organs are part of the reticuloendothelial system (RES), which is a component of the immune system. Its function is to remove antigens, such as proteinaceous nanoparticle structures, from circulation (Peiser *et al.*, 2002).

A potential advantage of plant VNPs over synthetic nanomaterials is that plant VNPs could be administered as edible vaccines or therapeutics. Plant VNPs are robust and can withstand harsh conditions. It has been shown that CPMV particles remain structurally sound under simulated gastric or intestinal conditions (simulated gastric or intestinal fluids containing proteases and acidic pH values of 2–6) (Rae *et al.*, 2005). Furthermore, CPMV particles have been shown to be bioavailable following oral administration in mice (Gonzalez *et al.*, 2009; Rae *et al.*, 2005).

Pharmacokinetic properties are highly dependent on surface charge and materials properties (reviewed in Li & Huang, 2008). The more positive the surface charge of a nanoparticle, the longer its plasma circulation time. This holds true for VNPs: the negatively charged VNPs CPMV and CCMV, for example, have rather short plasma circulation times of $t_{1/2} < 15$ min (Kaiser *et al.*, 2007; Singh *et al.*, 2007), whereas positively charged Qβ particles reach plasma half-life times of greater than 3 h (Prasuhn *et al.*, 2008). CPMV and CCMV particles display negative surface charges of −180e and −120e, respectively. In stark contrast, Qβ particles have a positive surface charge of +300e [values are derived from the Virus Particle Explorer Database (VIPERdb); http://viperdb.scripps.edu].

VNPs with altered surface charges were compared, where surface charge alteration was accomplished using chemical or genetic modification methods. Data obtained with each platform tested, Qβ, λ phage,[1] and M13, were in good agreement: the more positively charged the VNP, the longer its circulation time (Molenaar *et al.*, 2002; Prasuhn *et al.*, 2008; Srivastava *et al.*, 2004; Vitiello *et al.*, 2005). For example, non-modified M13 and Qβ particles remained circulating in plasma for several hours. Chemical modification at solvent-exposed, positively charged Lys side chains increased the net negative charge and greatly reduced plasma half-lives (Molenaar *et al.*, 2002; Prasuhn *et al.*, 2008). Many modification chemistries involve the covalent modification of Lys side chains on the surface of VNPs, and thus may alter the surface charge. This of course depends on the ligand attached; if a neutral or negatively charged molecule is attached to a Lys side chain, the overall surface charge of the VNP is expected to be more negative, thus leading to shorter circulation times. However, if a positively charged molecule is covalently linked to surface Lys residues, the overall surface

---

[1] λ phage is a DNA-containing coliphage with a head–tail structure; the head has $T = 7$ symmetry [Database of the International Committee on taxonomy on Viruses (ICTV); http://www.ncbi.nlm.nih.gov/ICTVdb].

charge should not change dramatically. This phenomenon has been found to hold true for other proteins, liposomes, and nano- and microspheres as well (Levchenko *et al.*, 2002; Moore *et al.*, 1977; Roser *et al.*, 1998; Wachsmuth & Klingmuller, 1978; Yamasaki *et al.*, 2002).

The impact on plasma circulation is not solely dependent on the surface charge; materials properties derived from the ligands attached will also impact the *in vivo* properties. For example, if PEG chains are attached to Lys side chains, the properties of the PEG typically outweigh the charge effect and longer circulation times are achieved (see Section 8.1.2). It is difficult to precisely predict the *in vivo* properties of a particular formulation; each formulation generated must be evaluated in detail prior to its clinical use.

8.1.2 PEGylation – A Strategy to Reduce Biospecific Interactions and Immunogenicity, and Increase Plasma Circulation Time

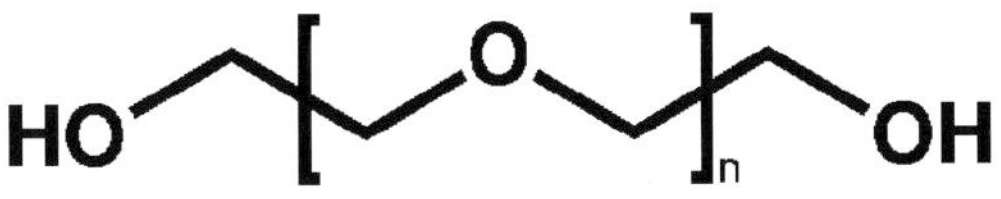

Figure 8.1 Structure of polyethylene glycol (PEG).

For certain applications in nanomedicine such as imaging or drug delivery, immunogenicity may be regarded as an undesired side effect. Strategies that overcome or reduce the immunogenic properties of VNPs have been developed; the most prominent technique is the attachment of polyethylene glycol (PEG) to biomolecules or nanomaterials, termed *PEGylation*.

Polythylene glycol (Fig. 8.1) is a non-charged, highly hydrophilic polymer. PEG is non-toxic and has been approved for use in humans by the U.S. Food & Drug Administration (FDA). Increasingly being used for pharmaceuticals and other biomedical applications, PEGylation efficiently reduces or blocks biospecific interactions, increases solubility and stability, increases plasma circulation time, and reduces immunogenicity. PEGylation has been applied to a variety of nanomaterials including liposomes, carbon nanotubes, and dendrimers (reviewed in Harris & Chess, 2003; Roberts *et al.*, 2002; Wattendorf & Merkle, 2008).

PEGylation can be achieved using a range of bioconjugation methods (see Chapter 4). PEG chains may be activated with a range of functionalities to allow covalent coupling to the desired molecule or material. Modified PEG chains are commercially available, as well as bivalent PEG molecules that allow interlinking of different functionalities.

PEG chains have been covalently attached to *Potato virus X* (PVX), *Tobacco mosaic virus* (TMV), and MS2 (Bruckman *et al.*, 2008; Kovacs *et al.*, 2007; Steinmetz *et al.*, 2009c) using NHS chemistries, click reactions, or oxime ligation (recall Chapter 4). PEGylation has been extensively studied using various CPMV formulations (Destito *et al.*, 2007; Lewis *et al.*, 2006; Raja *et al.*, 2003; Steinmetz & Manchester, 2009). PEGylation of CPMV before administration effectively shielded the particles from inducing a primary immune response (Raja *et al.*, 2003). PEGylating CPMV is also an effective strategy to avoid CPMV–cell interactions *in vitro* and *in vivo* (Destito *et al.*, 2007; Lewis *et al.*, 2006; Steinmetz & Manchester, 2009). It was shown that attaching as few as 30 PEG chains of a molecular weight of 2000 Da (this is a PEG chain consisting of $n = 14$ PEG monomers; see Fig. 8.1) to CPMV surface Lys side chains allowed effective shielding *in vitro* using cells and *ex vivo* using tumor tissues (Steinmetz & Manchester, 2009). Only 10% of the available surface Lys groups were modified with PEG [CPMV displays 300 addressable Lys side chains (Chatterji *et al.*, 2004)]. Modeling suggested that less than 1% of the available surface area of the VNP was covered and effectively shielded (Steinmetz & Manchester, 2009). Similar data were obtained studying PEGylated PVX formulations (Steinmetz *et al.*, 2009c). This implies that limited PEGylation of VNPs is sufficient for shielding, and thus a large surface area and many attachment sites remain available for further modification with targeting, imaging, and therapeutic ligands.

8.1.2.1 Stealth effect of PEGylated Ads and AAV for gene therapy

Ad vectors and AAV have been developed for gene therapy for many years. Broad clinical utility has been challenging because of (i) rapid clearance (the plasma half-life of Ad is less than 2 min), (ii) liver toxicity, and (iii) prevalence of Ad-neutralizing antibodies in an individual's serum (reviewed in Eto *et al.*, 2008).

A variety of PEGylated Ad vectors have been analyzed *in vivo*, and PEG attachment was beneficial for addressing each of the above-mentioned points: (i) PEGylated Ads had a fourfold higher clearance ratio compared with native Ads; (ii) enhanced tumor uptake and lower uptake in the liver implied reduced side effects and increased therapeutic success; and (iii) evasion of antibody neutralization by PEG-Ad formulations has been reported (reviewed in Eto *et al.*, 2008).

Careful design and evaluation of a particular formulation is critical, as shown in a study using AAV. It was found that there is only a small window of effective PEGylation of AAV, in which the virus remained fully infective allowing for efficient gene delivery while benefiting from a stealth effect (Lee *et al.*, 2005). The more the PEG chains and the larger the PEG chains attached, the better the shielding and stealth effect. On the downside, shielding is not exclusive to the components of the immune system. As AAV vectors rely on the native capsid specificity to interact with their target cells, if decorated too heavily with PEG chains AAV loses its ability to transfect cells (Lee *et al.*, 2005).

8.1.2.2 Stealth effect via genetic alteration

Studies focusing on AAV showed that a stealth effect could also be achieved by alterations of the particle surface. Mutant particles with altered antigenic surface properties have been generated using genetic modification protocols (Huttner *et al.*, 2003). However, compared with chemical modification protocols these methods can be cumbersome and time-consuming. Mutants with up to 70% reduced affinity to AAV-specific antibodies demonstrated antibody evasion *in vitro*. On the downside, the mutants suffered from significant losses in infectivity (Huttner *et al.*, 2003). This example also underlines the importance in carefully evaluating each formulation. It is critical to find the right balance between reducing undesired side effects while still maintaining the desired function of the VNP.

8.1.2.3 Dual-Functionalized Formulations That Facilitate Shielding While Maintaining Efficient Gene Delivery

To overcome the problems of PEGylated formulations being shielded from their target cells and tissues, targeting ligands that bind to specific molecular receptors have been utilized (discussed in detail in Sections 8.5 and 8.6). A series of PEG-Ad-RGD vectors have been developed and tested. The RGD (Arg-Gly-Glu) motif is the targeting ligand that promotes binding and internalization of the vectors into tumor cells (Sections 8.5 and 8.6). The RGD motif is highly specific for the integrins $\alpha_v\beta_3$ and $\alpha_v\beta_5$ overexpressed on tumor endothelium and cancer cells. Targeting these tumor markers has proven successful using a range of platforms and is envisaged as a promising strategy for imaging and drug delivery (Li *et al.*, 2003; Meyer *et al.*, 2006; Sipkins *et al.*, 1998; Winter *et al.*, 2003). PEG-Ad-RGD vectors have been shown to retain antibody evasion while maintaining efficient gene delivery rates (reviewed in Eto *et al.*, 2008). The principles are outlined in Fig. 8.2.

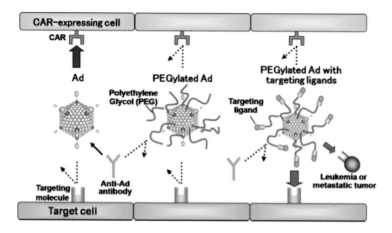

Figure 8.2 Characteristics of PEGylated Ads. Non-modified Ads (left panel) bind to cells expressing the coxackie adenovirus receptor (CAR) and are recognized by neutralizing antibodies. PEGylated Ads (middle panel) do not bind to cells expressing CAR and are not recognized by neutralizing antibodies. These Ads, however, also do not bind to target cells. PEGylated Ads displaying targeting ligands on the periphery of the PEG chains are shielded from CAR-expressing cells and antibodies. Targeting facilitates uptake by target cells and gene delivery. Reproduced with permission from Eto, Y., Yoshioka, Y., Mukai, Y., Okada, N., and Nakagawa, S. (2008) Development of PEGylated adenovirus vector with targeting ligand, *Int. J. Pharm.*, **354**(1–2), 3–8.

8.2 VACCINE DEVELOPMENT USING VNPs

Because of their highly repetitive molecular structures, VNPs have varying degrees of immunogenicity (Acosta-Ramirez *et al.*, 2008; Kaiser *et al.*, 2007; Lacasse *et al.*, 2008; Raja *et al.*, 2003; Smith *et al.*, 2006, 2007). For vaccine development, immunogenicity of the carrier is beneficial as it helps to boost the immune response.

Several classes of anti-viral vaccines have been developed for use in humans and animals. These include inactivated virus vaccines, attenuated virus vaccines, VLPs, and chimeric VLPs/VNPs (reviewed in Garcea & Gissmann, 2004; Grgacic & Anderson, 2006; Ludwig & Wagner, 2007). The first anti-viral vaccines were based on *inactivated* viruses or *attenuated* strains. Chemicals such as formalin can be used to generate an *inactivated* virus vaccine, whereby the virus does not remain infectious and is thus safe to use. However, some of the structural or chemical properties of the virus may be altered as a result of chemical treatment. Alteration or masking (or blocking) of antigenic surface epitopes can result in inefficiency or failure of the vaccine, or occasionally lead to induction of an adverse response.

An *attenuated* virus strain is a strain whose pathogenicity has been reduced. Attenuated virus strains are still infectious; however, their virulence is reduced to a level where it is no longer dangerous. Attenuated viruses have historically been generated by serial passage through non-human hosts or cell lines, and through such passage undergo selection for genetic modifications that result in reduced virulence in humans. The overall structural and immunogenic properties of attenuated viruses are usually not sufficiently altered and provide natural interactions with cells and the immune system. Some disadvantages of attenuated virus vaccines are when the viruses cannot be sufficiently attenuated to eliminate adverse affects while maintaining effective immune priming, as well as potential reversion to the virulent genotype (Doan *et al.*, 2005).

A more recent trend is the development of VLPs as vaccines. VLPs are self-assembled from coat protein monomers. The monomers can be produced in heterologous expression systems; a range of systems are available as discussed in Chapter 3. Purified coat proteins can be self-assembled into intact nucleic acid-free VLPs. VLPs do not contain viral nucleic acids, and are thus non-infectious and replication-deficient, making them safer for use in humans.

A number of VLP-based vaccines have been licensed, or are in clinical and preclinical trials (reviewed in Garcea & Gissmann, 2004; Grgacic & Anderson, 2006; Ludwig & Wagner, 2007). Commercialized VLP-based vaccines have been shown to elicit a protective immune response in humans; for example, VLP-based vaccines using VLPs formed by human papilloma virus (HPV), polyomavirus, or hepatitis B virus (HBV) coat proteins have been shown to be protective against the pathogens (reviewed in Garcea & Gissmann, 2004; Grgacic & Anderson, 2006; Ludwig & Wagner, 2007).

Chimeras are VLP- or VNP-based assemblies that deliver immunogenic epitopes of other pathogens (an epitope is an immunogenic peptide or protein fragment). HBV VLPs and HPV VLPs, for example, have been used as platforms for the presentation of foreign (= non-HBV- or HPV-derived) antigenic sequences. The VLP is used as a carrier and serves as an adjuvant to boost the immune response. Besides utilizing VLPs derived from human pathogens, a popular strategy exploits the use of plant VNPs and phages as epitope carriers.

VLPs and VNPs are immunogenic because of their highly ordered and repetitive structures that are recognized as pathogen-associated molecular patterns (PAMPs) by the immune system. The quasi-crystalline nature of the VLP/VNP facilitates pattern recognition by specific and innate immune responses. The highly immunogenic properties of VLPs and VNPs are derived from their interaction with and activation of dendritic cells.

VLPs have also been reported to interact directly with macrophages and monocytes, further stimulating induction of the immune response (reviewed in Garcea & Gissmann, 2004; Grgacic & Anderson, 2006; Ludwig & Wagner, 2007).

Chimeric VNPs displaying antigenic peptide epitopes on the exterior surface have been developed, where the viral capsid is used as a scaffold for presenting the epitope and also serves as an adjuvant to boost the immune response (Acosta-Ramirez *et al.*, 2008; Cornuz *et al.*, 2008; Kaiser *et al.*, 2007; Lacasse *et al.*, 2008; Manayani *et al.*, 2007; Maurer *et al.*, 2005; Raja *et al.*, 2003; Smith *et al.*, 2006, 2007). A broad range of plant VNPs have been developed as platforms for display of immunogenic epitopes. These include the plant viruses *Cucumber mosaic virus* (CMV), CPMV, RCNMV, *Tomato bushy stunt virus* (TBSV), PVX, and TMV (reviewed in Canizares *et al.*, 2005; Johnson *et al.*, 1997; Porta & Lomonossoff, 1998; Yusibov *et al.*, 2006). For example, CPMV has been used for the display of the *Canine parvovirus* VP2 epitope. Immunized dogs were protected against lethal challenge with the pathogen (Langeveld *et al.*, 2001; Nicholas *et al.*, 2002). Similar data were obtained in mink immunized with CPMV particles displaying the *Mink enteritis virus* VP2 epitope. The animals were protected against viral challenge (Dalsgaard *et al.*, 1997). Heterologous PVX particles displaying the HIV type 1 ELDKWA epitope caused the production of neutralizing antibodies in mice (Brennan *et al.*, 1999). Some examples of TMV epitope display used epitopes from murine hepatitis virus or *Foot and mouth disease virus* (FMDV), and were successful in conferring protection against virulent challenge in mice or guinea pigs, respectively (Koo *et al.*, 1999; Wu *et al.*, 2003).

Phages and the insect virus *Flock House virus* (FHV) (see below) have also been developed as platforms for potential vaccines. A potential vaccine against nicotine has been recently developed based on the bacteriophage Qβ. This candidate vaccine is envisioned as a promising strategy for smoking cessation. By inducing neutralizing antibodies against the addictive compound nicotine, and reducing blood nicotine levels thereby limiting transport across the blood–brain barrier, a nicotine vaccine is expected to reduce cigarette addiction. Qβ particles covalently modified with nicotine are currently under investigation in clinical trials (Cornuz *et al.*, 2008; Maurer *et al.*, 2005).

Recombinant FHV particles have been developed that have the potential to function as an anthrax antitoxin and vaccine. Anthrax is an acute and highly lethal disease caused by the bacterium *Bacillus anthracis*. Based on the potential use of anthrax as a weapon of bioterrorism, there is need for the development of improved antitoxins and vaccines. Morbidity and mortality from anthrax are caused mainly by the action of anthrax toxin, which consists of a cell binding protein, protective antigen (PA), and two enzymatic components, edema factor and lethal factor. During infection the toxin components are produced by *B. anthracis* and PA binds to anthrax toxin receptors, which are widely distributed on human cells, and leads to internalization of the toxin (Mock & Fouet, 2001; Mock & Mignot, 2003). FHV particles were genetically engineered to express and display the anthrax receptor on the surface of the viral capsid (Fig. 8.3a) (Manayani *et al.*, 2007). These anthrax receptor-displaying FHV particles efficiently bind to PA. Studies in a rat animal model showed that this strategy inhibited lethal toxin action (Manayani *et al.*, 2007). Furthermore, it was demonstrated that FHV-receptor chimeras further complexed with PA produced a multivalent array of PA antigen (Fig. 8.3b). This complex was more immunogenic than monomeric PA, and successfully elicited the production of antibodies that protected rats from anthrax lethal toxin challenge after a single dose (Manayani *et al.*, 2007).

In summary, a variety of VNPs have been utilized as platforms for the development of novel vaccines. The highly ordered protein scaffold in combination with the ease of manipulation (genetic or chemical engineering) makes VNPs attractive candidates for such development.

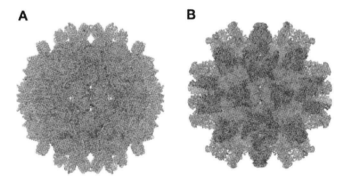

Figure 8.3 3-D models of FHV-VWAANTXR2 VLPs (anthrax-receptor-displaying *Flock House virus* particles) alone (a) or with bound PA83 (PA = protective antigen) (b). Pseudoatomic models of the FHV-VWAANTXR2 chimera. X-ray coordinates of FHV capsid protein (green) and ANTXR2 VWA domain (yellow) were docked into the cryo-electron microscopic density. Surface views of the particles in the absence of the cryoEM density maps. (b) *In silico* model of PA83 bound to the surface of FHV-VWAANTXR2 chimeras. PA83 (purple) was modeled onto the surface of the FHV-VWAANTXR2 VLP using the known high-resolution X-ray structure of the ANTXR2-VWA/PA63 complex as a guide. Reproduced Manayani, D. J., Thomas, D., Dryden, K. A., Reddy, V., Siladi, M. E., Marlett, J. M., Rainey, G. J., Pique, M. E., Scobie, H. M., Yeager, M., Young, J. A., Manchester, M., and Schneemann, A. (2007) A viral nanoparticle with dual function as an anthrax antitoxin and vaccine, *PLoS Pathog.*, **3**(10), 1422–1431.

8.3 VNPs AS SENSORS FOR BIOMEDICAL IMAGING

Various imaging techniques have been developed for non-invasive *in vivo* imaging, such as computed tomography (CT), magnetic resonance imaging (MRI, see Section 8.3.2), positron emission tomography (PET), as well as optical imaging (Section 8.3.1) (reviewed in Cai & Chen, 2007; Graves *et al.*, 2004; Rudin & Weissleder, 2003). Early detection, treatment, and monitoring of disease have great potential to improve patient outcomes. Furthermore, development and improvement of imaging modalities and techniques play a key role in biomedical research. The development of ligands for tumor-specific markers as well as vascular homing peptides (Arap *et al.*, 2002; Hajitou *et al.*, 2006a; Nanda & St. Croix, 2004; Ruoslahti, 2002) has opened the door for creating novel targeted imaging devices with higher specificity for desired locations and reduced background in the surrounding tissues.

VNPs and other nanomaterials can be utilized to interlink targeting and imaging modalities. VNPs have been combined with organic fluorophores, QDs, metallic nanoparticles, and gadolinium (Gd) complexes (reviewed in Manchester & Steinmetz, 2008; Singh *et al.*, 2006b; Steinmetz & Evans, 2007; Steinmetz *et al.*, 2009b; Young *et al.*, 2008). These hybrid VNPs can be fabricated using a broad range of design principles, including bioconjugation (Chapter 4), encapsulation, and infusion (Chapter 5), or by making use of biomineralization techniques (Chapter 6).

Hybrid VNPs can be categorized into three main groups:

Fluorescent VNP hybrids: These are VNPs covalently modified with organic biodegradable dyes for applications in fluorescence optical imaging. The feasibility of intravital vascular imaging using fluorescent dye-labeled CPMV probes has been demonstrated (see Section 8.3.1).

Lanthanide–VNP complexes: These are VNPs displaying lanthanides, such as Gd^{3+} or Tb^{3+}. Applications of such complexes are envisioned in the field of MRI imaging (see Section 8.3.2).

VNP–inorganic nanoparticle complexes: These are VNPs decorated with or encapsulating inorganic nanoparticles. Mineralized or metalized structures also fall into this category. VNPs with metal or QD cores can be regarded as plasmonic composite materials with potentially interesting biosensing applications. These materials can be activated using a light source; plasmons (i.e., density waves of electrons) are generated upon activation that can be detected using various spectroscopy methods (Section 8.3.3).

8.3.1 Fluorescent-Labeled VNPs for Intravital Vascular Imaging

The feasibility of fluorescent-labeled CPMV particles for intravital vascular imaging in live mouse embryos has been demonstrated (Fig. 8.4)

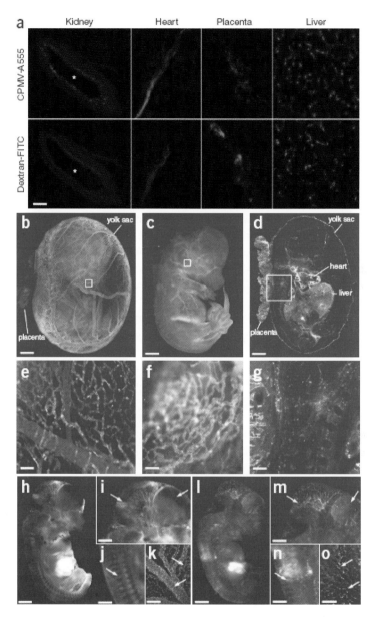

Figure 8.4 Fluorescent dye-conjugated *Cowpea mosaic virus* particles (CPMV-A555) enable visualization of vasculature intravitally and in fixed tissues. (a) Fluorescence images of tissue cryosections from kidney (asterisk indicates vessel lumen), heart, placenta, and liver isolated from adult mice coinjected with CPMV-A555 (top panels) and fluorescein dextran (bottom panels). Scale bar, 50 μm. (b, c) Intravital imaging of CPMV-A555 perfused 11.5-day embryo with the yolk sac intact (b) and removed (c). White boxes indicate the regions magnified in (e) and (f). Scale bar, 1.1 mm. Images

were captured approximately 1 h after injection. (d) Cryosection of an 11.5-day mouse embryo perfused with CPMV-A555. White box indicates the region magnified in (g). Scale bar, 1.1 mm. (e) Yolk sac vasculature, magnified. Scale bar, 25 µm. (f) Capillaries in the head region. Scale bar, 25 µm. (g) Intersomitic and placental vessels in embryo tissue section. Scale bar, 100 µm. (h–o) Comparison of intravital imaging with CPMV-A555 (h–k) and fluorescent nanospheres (e–o) in E11.5 mouse embryo. (h, l) Whole embryo. Scale bar, 1.1 mm. (i, m) Head region, arrows indicate areas of differential staining of anterior vasculature, brain vasculature. Scale bar, 770 µm. (j, n) CPMV staining allows increased resolution of intersomitic vessels. Scale bar, 540 µm. (k, o) Arrows indicate capillary and larger vessels of yolk sac membrane. Scale bar, 50 µm. Reproduced Lewis, J. D., Destito, G., Zijlstra, A., Gonzalez, M. J., Quigley, J. P., Manchester, M., and Stuhlmann, H. (2006) Viral nanoparticles as tools for intravital vascular imaging, *Nat. Med.*, **12**(3), 354–360.

(Lewis *et al.*, 2006). Fluorescent-labeled CPMV particles were injected intravenously in chick and mouse embryos, and imaging was performed over a time frame of 72 h. CPMV was internalized by endothelial cells and could be detected lining the vasculature, thus allowing high-resolution imaging. Major blood vessels as well as microvasculature could be visualized to a depth of up to 500 µm. Such high specificity and resolution could not be achieved when using fluorescent-labeled nanospheres, which are a current state-of-the-art imaging platform used in optical imaging (Fig. 8.5) (Lewis *et al.*, 2006).

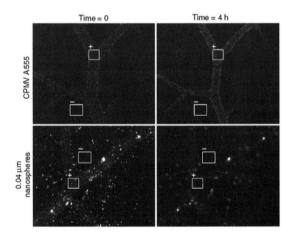

Figure 8.5 Comparison of intravital vascular staining intensity over time in the chick embryo. Comparative analysis of changes in vascular staining intensity over time in the chick embryo using fluorescent-labeled *Cowpea mosaic virus* (CPMV) particles or 0.04 µm nanospheres. Representative images captured immediately after injection (time = 0) and 4 h after injection (time = 4 h) showing sampled regions of interest (ROI) during intravital retained fluorescence assay. Fluorescence average of negative ROI (−) was subtracted from positive ROI (+). Reproduced from Lewis, J. D., Destito, G., Zijlstra, A., Gonzalez, M. J., Quigley, J. P., Manchester, M., and Stuhlmann, H. (2006) Viral nanoparticles as tools for intravital vascular imaging, *Nat. Med.*, **12**(3), 354–360.

Combined with targeting molecules to achieve destination-specific imaging these VNPs would be a powerful tool for non-invasive detection and visualization of disease and progression. Furthermore, optimization of the system, such as the use of near-infrared (NIR) probes, is expected to improve sensitivity. The feasibility of using CPMV for NIR fluorescence tomography has also been demonstrated (Wu *et al.*, 2005).

8.3.2 VNPs as Contrast Agents for MRI

MRI provides a powerful non-invasive imaging tool and is one of the most frequently utilized techniques for *in vivo* imaging. MRI is based on the alignment of protons from hydrogen atoms in a strong magnetic field. The aligned protons are then exposed to a pulse of radio waves, which leads to absorption of energy. When the second field is turned off, energy is released and can be detected by the scanner to generate the MRI image. Contrast agents are typically used to increase the brightness of the image and hence sensitivity of the technique. The principle of contrast agents is to increase relaxation time of water protons, the process by which the nuclear magnetization (alignment of protons) returns to equilibrium distribution. Gadolinium (Gd^{3+}) is commonly used as a contrast agent (reviewed in Aime *et al.*, 2002). Enhanced relaxivities can be achieved by coupling the contrast agent (Gd^{3+}) to a macromolecular carrier such as dendrimers or liposomes (Immordino *et al.*, 2006; Kobayashi & Brechbiel, 2005).

More recently, VNPs including CPMV, CCMV, MS2, and Qβ have been found to be promising platforms for contrast agent development (reviewed in Liepold *et al.*, 2007). VNPs are rigid structures with large rotational correlation times, resulting in increased relaxivity. In addition, because of the polyvalent nature of VNPs several hundred Gd^{3+} molecules can be complexed or attached to the VNP. It has been shown that Gd^{3+} molecules can be covalently attached to the exterior or interior surface, complexed with encapsidated RNA molecules, or bound at intrinsic metal-binding sites at coat protein interfaces (Allen *et al.*, 2005; Anderson *et al.*, 2006; Hooker *et al.*, 2007; Prasuhn *et al.*, 2007) (recall Chapters 4 and 5 with regard to the attachment strategies). Each of these paramagnetic VNP formulations exhibited extraordinarily high relaxivities, indicating that VNPs could serve as excellent candidates for MRI contrast agents. An *in vivo* evaluation of the performance of the materials has not yet been reported.

8.3.3 VNP–Inorganic Nanoparticle Complexes

A large variety of inorganic VNP hybrids have been developed, all of which have potential for biomedical imaging applications and therapeutic applications (this section and Section 8.4). To date, most of these materials have been characterized biochemically and biophysically. Only a very few examples describe an evaluation of the materials in cell culture (see below). *In vivo* data have not been reported yet.

As described in Chapter 5, the self-assembly mechanisms of VNPs or VLPs can be exploited to encapsulate synthetic nanoparticles inside the VNP or VLPs. The utility of SV40 VLPs-encapsulating QDs for live imaging to follow virus internalization and trafficking has been demonstrated (Fig. 5.11) (Li *et al.*, 2009). The advantage of incorporating the imaging moieties such as QDs on the inside of the VLP is that the exterior surface remains unmodified. The exterior surface is involved in receptor recognition and internalization so modifications can lead to abrogation of target specificity and cell entry. This has been observed using a chemically engineered Ad vector (Everts *et al.*, 2006). Ad vectors were covalently modified with gold nanoparticles; the goal was to engineer re-targeted gold nanoparticle-modified Ads for a combination of hyperthermia treatment (cell killing via heat) and gene therapy. Gold nanoparticles were attached using NHS coupling to surface Lys side chains. Infectivity and efficient gene delivery was only retained at low ratio labeling with gold nanoparticles. This, however, limits the potential for efficient hyperthermia effects (Everts *et al.*, 2006). This strategy could be improved by fine-tuning and optimizing the chemistry and testing various attachment sites or by incorporating the gold nanoparticles into the interior of the vector (Everts *et al.*, 2006).

In a different example, hybrid hydrogels consisting of M13 and gold nanoparticles were investigated for cell sensing applications. Biocompatible and tunable networks consisting of chimeric M13 particles displaying cell-binding peptides and gold nanoparticles were formed on the basis of electrostatic interactions. Cell binding and receptor-mediated internalization of the phage particles, even when incorporated into the hybrid network, were facilitated (Souza *et al.*, 2007).

When developing hybrid systems that combine VNPs with synthetic materials, it is important to re-evaluate the pharmacological and potential toxic side effects. For example, recent biological studies revealed significant cytotoxicity of QD nanocrystals (Portney & Ozkan, 2006). Similar data have been reported for carbon nanotubes; because of their non-biodegradable nature, cellular persistence and cytotoxic effects can occur with these particles (Hardman, 2006; Prato *et al.*, 2008; Takagi *et al.*, 2008). These are important factors to be considered when working with VNP-inorganic hybrids.

8.4 TARGETING VNPs TO SITES OF DISEASE

Targeting imaging modalities and therapeutic molecules specifically to sites of disease while avoiding healthy tissues is an important goal in biomedicine. Destination-specific delivery of imaging or therapeutic molecules will facilitate greater specificity and organ avoidance and therefore limit undesired side effects. The development of phage display technologies has led to the identification of tumor-specific markers and their ligands, as well as vascular homing peptides (Arap *et al.*, 2002; Hajitou *et al.*, 2006a; Nanda & St. Croix, 2004; Ruoslahti, 2002). These ligands specifically recognize receptors overexpressed on diseased tissues. The discovery of these ligands has revolutionized the field and opened the door for creating specifically targeted devices (Section 8.4.2).

8.4.1 Natural VNP–Cell Interactions

8.4.1.1 CPMV—A VNP with Natural Affinity for Mammalian Endothelial Cells

Intravital imaging studies using fluorescent-labeled CPMV particles indicated that CPMV particles were specifically internalized by endothelial cells *in vivo* (Lewis *et al.*, 2006). This interaction has been dissected in detail; it was shown that this interaction is biospecific and mediated by the mammalian protein vimentin (Koudelka *et al.*, 2007, 2009). Vimentin is a type III intermediate filament predominantly expressed in the cytosol of cells of mesenchymal origin. Cytosolic vimentin plays a key role in intracellular dynamics and archtitecture (reviewed in Evans, 1998; Wang & Stamenovic, 2002). Besides its cytosolic localization, vimentin has recently been identified as a surface-exposed and/or secreted protein in activated macrophages (Mor-Vaknin *et al.*, 2003), T-lymphocytes (Huet *et al.*, 2006; Nieminen *et al.*, 2006), endothelial cells, specifically in tumor tissue (van Beijnum *et al.*, 2006), and endothelial venules of lymph nodes (Xu *et al.*, 2004), among others. Vimentin plays a role in situations such as tumor development and progression (reviewed in Brabletz *et al.*, 2005; Gilles *et al.*, 2003; Kokkinos *et al.*, 2007).

    The fact that the plant pathogen specifically interacts with a mammalian protein may be explained by the structural similarities that CPMV shares with other members of the picornavirus superfamily. *Theiler's murine encephalomyolitis virus* (TMEV), also a member of the picornavirus superfamily, shows similarities in genetic organization and structure to CPMV. TMEV has also been shown to directly interact with vimentin (Nedellec *et al.*, 1998).

It has been suggested that surface vimentin is overexpressed in tumor endothelium (van Beijnum *et al.*, 2006); this correlates with high uptake of CPMV in tumor endothelium as shown in studies using the chick choreoallantoic membrane tumor model. Fluorescent-labeled CPMV sensors accumulated within tumor endothelium and allowed imaging and mapping over long time periods (Fig. 8.6) (Lewis *et al.*, 2006). The use of CPMV as a natural endothelial probe is envisioned in imaging vascular disease and may provide novel insights into the expression pattern of surface vimentin.

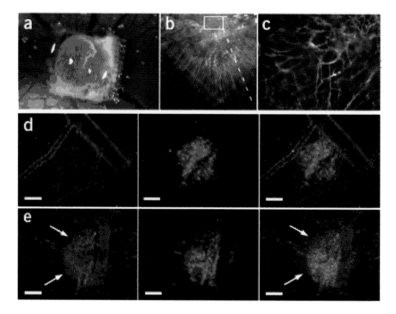

Figure 8.6 Fluorescent *Cowpea mosaic virus* (CPMV) nanoparticles highlight tumor angiogenesis: intravital imaging in a CAM/HT1080 fibrosarcoma model. (a) Bright-field image of HT1080 tumor on-plant on the chick chorioallantoic membrane (CAM) at 7 d (d = days). Opaque object is a nylon mesh grid used for quantification of angiogenesis. (b) Fluorescence image of tumor on-plant after injection of embryo CPMV-A555. (c) High-magnification image of tumor interior shown in (b); tumor microvasculature is clearly visualized . (d, e) Visualization of HT1080 tumor angiogenesis using CPMV-A555. (d) Left, visualization of pre-existing vasculature in the CAM immediately after HT1080 tumor cell injection using CPMV-A555. Middle, GFP-expressing HT1080 tumor bolus under the surface of the CAM. Right, merge. Scale bar, 100 mm . (e) Left, visualization of pre-existing CAM vasculature and neovasculature arising from tumor angiogenesis 24 h after tumor cell injection. Middle, GFP-expressing HT1080 tumor bolus. The extensive migration over 24 h indicates a high level of tumor cell viability. Right, merge. Scale bar, 100 mm. Reproduced from Lewis, J. D., Destito, G., Zijlstra, A., Gonzalez, M. J., Quigley, J. P., Manchester, M., and Stuhlmann, H. (2006) Viral nanoparticles as tools for intravital vascular imaging, *Nat. Med.*, **12**(3), 354–360.

The natural interactions of CPMV with the mammalian system were further exploited, and targeting to sites of inflammation in the central nervous system (CNS) was demonstrated (Shriver *et al.*, 2009). In brief, mice infected with neurotropic mouse hepatitis virus were studied; this is a commonly used animal model for the human demyelinating disease multiple sclerosis. CPMV was found to localize mainly to the CNS endothelium that contained an intact blood–brain barrier. In inflammatory lesions containing macrophage/microglial cell infiltration and IgG, which indicates breakdown of the blood–brain barrier, CPMV was also detected in the brain parenchyma (Shriver *et al.*, 2009). This may open the door for targeted treatment of inflammatory disease of CNS.

8.4.1.2 Designing VNPs for Targeting Applications

Some cells and tissues in diseased sites express a different set of receptors on their cell surface compared to cells in healthy tissues. Cells in tumor tissue are rapidly proliferating. Therefore, tumor cells overexpress certain proteins and receptors on the cell surface that are involved in signaling nutrient uptake, cell growth and division. As selectively and significantly overexpressed, these receptors provide a target for tissue-specific delivery of imaging modalities or therapeutics.

The receptor for the protein transferrin (Tf), for example, is overexpressed on several tumor cells. Tf is a circulating iron storage protein. Tf binds to the Tf receptors (TfR) and is internalized. Inside the cells, iron is released and Tf recycled. Iron is in high demand during cell growth and cell proliferation; hence, proliferating tumor cells overexpress TfR on their cell surface. Normal, non-dividing cells express few to negligible numbers of TfR on their cell surface. In stark contrast, tumor cells express 10^5 or more TfR per cell. Targeting TfR is thus expected to be a promising strategy to specifically target cancer (Becker *et al.*, 2000; Bridges & Smith, 1985; Du *et al.*, 2002; Gomme *et al.*, 2005; Inoue *et al.*, 1993; Li & Qian, 2002; Li *et al.*, 2002; Qian *et al.*, 2002; Ryschich *et al.*, 2004; Sato *et al.*, 2000).

Delivering VNPs to tumor cells exploiting TfR has been studied using MS2, CPMV, and CPV (Brown *et al.*, 2002; Sen Gupta *et al.*, 2005; Singh *et al.*, 2006a). For MS2 and CPMV, the protein Tf has been covalently attached and displayed on the particle surface (Brown *et al.*, 2002; Sen Gupta *et al.*, 2005); this was achieved using the principles discussed in Chapter 4. Tf-targeted MS2 particles were further modified with therapeutic molecules and targeted drug delivery was demonstrated (Brown *et al.*, 2002) (see Section 8.5). TfR targeting can be achieved making use of the natural receptor-binding properties of CPV, which utilizes TfR in order to infect cells. Non-infectious and non-replicating CPV VLPs were developed that preserve binding to TfR (Singh *et al.*, 2006a). Delivery of dye-labeled CPV VLPs to different cancer cell lines *in vitro* has been demonstrated (Fig. 8.7) (Singh *et al.*, 2006a).

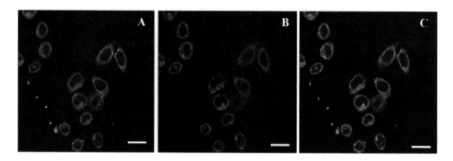

Figure 8.7 Binding and internalization of *Canine parvovirus* (CPV) VLPs into HeLa cells (human cervical cancer cell line). HeLa cells incubated with Texas red–labeled transferrin (red) and CPV VLPs were washed and fixed. Labeled antibodies (green) were used to detect the presence of CPV VLPs in the cells by fluorescence confocal microscopy. (a) CPV VLPs are seen as green areas in the cytoplasm, (b) localization of Texas red-transferrin (red) and (c) merged picture showing co-localization of CPV-VLPs and transferrin in yellow. Scale bar, 25 µm. Reproduced from Singh, P., Destito, G., Schneemann, A., and Manchester, M. (2006) Canine parvovirus-like particles, a novel nanomaterial for tumor targeting, *J. Nanobiotechnol.*, 4, 2.

Another receptor that is overexpressed specifically on tumor cells is the receptor for folic acid (FA). FA is a vitamin required during growth and development. Proliferating cells require high levels of this vitamin and therefore overexpress the FA receptor (FR). Targeting nanomaterials to FR has been shown to be a promising strategy (Cafolla *et al.*, 2002; Chen *et al.*, 1998; Leamon & Low, 1992, 1993, 2002a,b; Reddy & Low, 1998; Turek *et al.*, 1993; Wang & Low, 1998). The VNPs CPMV and *Hibiscus chlorotic ringspot virus* (HCRSV) have been covalently modified with FA. Multivalent display of FA on the particle surface facilitated cell internalization by tumor cells overexpressing FR (Destito *et al.*, 2007; Ren *et al.*, 2007). In the case of CPMV, particles were covalently modified with PEG and FA; PEG provides shielding of the natural CPMV interactions, and FA facilitated target specificity (Fig. 8.8) (Destito *et al.*, 2007). The FR-targeting strategy has been further developed for targeted delivery of the chemotherapeutic doxorubicin using the HCRSV platform (see Section 8.5) (Ren *et al.*, 2007).

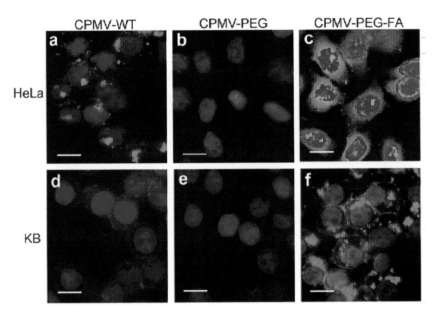

Figure 8.8 Fluorescent microscopy of monolayers of HeLa (top) and KB (bottom) cells. Cells were incubated with the indicated virus particles (37°C, 2 h), followed by permeabilization and treatment with anti-CPMV antibody and then with a green fluorescent secondary antibody conjugate. Nuclei are stained with DAPI (blue). Scale bar, 10 μm. Reproduced from Destito, G., Yeh, R., Rae, C. S., Finn, M. G., and Manchester, M. (2007) Folic acid-mediated targeting of cowpea mosaic virus particles to tumor cells, *Chem. Biol.,* **14**(10), 1152–1162.

The development of phage display technologies has led to the identification of tumor-specific markers and their ligands, as well as vascular homing peptides (Arap *et al.,* 2002; Hajitou *et al.,* 2006a; Nanda & St. Croix, 2004; Ruoslahti, 2002). A prominent ligand identified using this approach is the RGD motif. The RGD sequence is a popular targeting ligand used. RGD peptides recognize the integrins $\alpha_v\beta_3$ and $\alpha_v\beta_5$, which are overexpressed on tumor endothelium and cancer cells (Li *et al.,* 2003; Meyer *et al.,* 2006; Sipkins *et al.,* 1998; Winter *et al.,* 2003).

The RGD motif has been widely exploited in viral nanotechnology, and genetically engineered as well as chemically modified VNPs displaying RGD sequences on the particle surface have been developed and tested: RGD has been chemically attached to CPMV (Sen Gupta *et al.,* 2005), PEG and RGD were linked to TMV (Bruckman *et al.,* 2008), and the RGD sequence has been genetically engineered into M13 phages and AAV vectors (Hajitou *et al.*, 2006b; Pasqualini *et al.*, 1997; Souza *et al.*, 2006) (see also Section 8.6). *In vivo* experiments using mice bearing human breast tumor xenografts (axenograft is generated by transplanting tumor cells of a different species, typically via subcutaneous injection into the flank of the animal) confirmed tumor homing of M13 phages displaying the RGD motif (Pasqualini *et al.*, 1997).

The specific delivery of nanomaterials to sites of disease, such as cancer, opens the door to the development of "smart" nanodevices for targeted drug delivery (see below).

8.5 THERAPEUTIC APPROACHES

VNPs have been designed for therapeutic approaches such as the delivery of drugs (chemotherapy). In addition, efforts have been made toward alternative therapies including photothermal therapy (PTT) and photodynamic therapy (PDT). PTT and PDT are closely related. In PTT electromagnetic radiation, such as infrared, is applied to activate a sensitizer, to release vibrational energy (i.e., heat) to kill the target cells. In PDT, a photosoensitizer is excited with specific band light (or wavelength) to generate reactive oxygen species to kill target cells. PTT and PDT can be used for localized treatment. A limitation is that the treatment can only be applied to the surface of tissues and is dependent on efficient penetration of the radiation or light (Huang *et al.*, 2008; Ortel *et al.*, 2009).

A variety of strategies have been developed that allow generation of hybrid VNP systems that are combined with synthetic materials such as iron oxide, gold, and other metallic nanoparticles (recall Chapters 5 and 6), all of which have potential for PTT or PDT. Evaluation of these materials *in vitro* or *in vivo* has not been performed yet.

Besides using synthetic nanoparticles for such approaches, chemical complexes such as ruthenium or fullerenes have also been shown to be promising candidates for PDT applications. Derivatives of C_{60} ("Buckyball") have exceptional radical-scavenging properties and are as such good candidates for use as photosensitizers in PDT. A major drawback of fullerene material is its insolubility in aqueous solution. It was recently shown that the solubility of C_{60} can be significantly enhanced through conjugation and multivalent display using CPMV VNPs or Qβ VLPs. The VLP acts as a hydrophilic carrier allowing cell delivery of the material. Biochemical and biophysical data indicated multivalent display of around 40 C_{60} molecules per VNP. Indeed, *in vitro* studies confirmed the efficient delivery of the hybrid material into cells (Fig. 8.9) (Steinmetz *et al.*, 2009a). Whether photodynamic cell killing can be achieved has not yet been demonstrated.

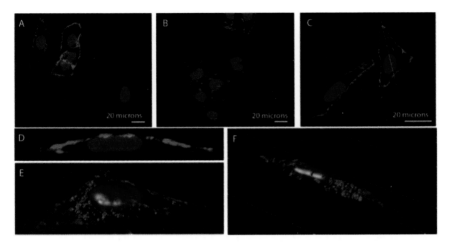

Figure 8.9 Confocal microscopy. (a) HeLa cells only. (b–f) Cells treated with Qβ-PEG-C60-A568 particles. Color key: blue, nuclei (DAPI); red, Qβ-PEG-C60-A568; green, A488-labeled wheat germ agglutinin. (d) Z-section image (1.2 μm deep) recorded along the line shown in (c); step size 0.3 μm. (e, f) Same cell as shown in (d), image reconstructions using Imaris software. Reproduced with permission from Steinmetz, N. F., Hong, V., Spoerke, E. D., Lu, P., Breitenkamp, K., Finn, M. G., and Manchester, M. (2009) Buckyballs meet viral nanoparticles: candidates for biomedicine, *J. Am. Chem. Soc.*, **131**(47), 17093–17095.

In a different approach, CCMV particles were dual-functionalized and targeted cell killing via PDT was demonstrated (Suci *et al.*, 2007b). In short, CCMV was covalently modified with ruthenium complexes, which serve as the photosensitizers. The complexes were specifically targeted to the pathogenic, biofilm-forming bacteria *Staphylococcus aureus*. This was performed by attachment of specific antibodies (anti-protein A). Biotinylated CCMV–ruthenium complexes and biotinylated antibodies were interlinked making use of streptavidin (recall Section 4.3.8). Targeting of the VNP sensors and specific binding to *S. aureus* was confirmed (Fig. 8.10) (Suci *et al.*, 2007a,b). When such VNP-targeted bacteria were exposed to light-emitting diodes (at a wavelength of 470 nm), cell killing was observed. This technology may lead to novel routes for antimicrobial PDT (Suci *et al.*, 2007b).

Phages such as M13 and fd (also a filamentous bacteriophage) have been developed for antimicrobial treatment of *S. aureus*. Targeting was achieved making use of specific antibodies. Cell killing was induced via delivery of the antibiotic chloramphenicol (Yacoby *et al.*, 2006).

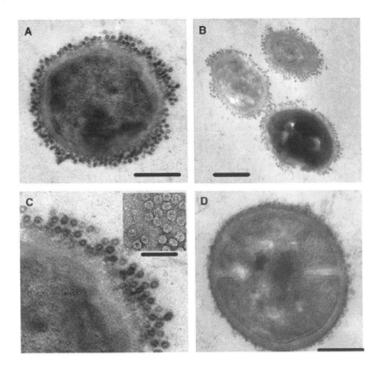

Figure 8.10 TEM thin sections showing high-density coverage of *Cowpea chlorotic mottle virus* (CCMV) targeted to *Staphylococcus aureus* cells (a, b). Biotinylated = CCMV-biotin (CCMV-B) bound to *S. aureus* cells via the StAv/anti-SpA mAb-B linkage: the linkage consist of a biotinylated *S. aureus*-specific antibody that is linked via streptavidin to the biotinylated CCMV particle. The scale bars in (a) and (b) are 200 and 400 nm, respectively. (c) Magnified view of image presented in (a) with an insert of a TEM image of CCMV-B adsorbed on Formvar presented at the same scale as the thin section. The scale bar is 100 nm. (d) Representative image of an *S. aureus* cell from the negative control indicating a negligible level of non-specific binding. The scale bar is 200 nm. Reproduced with permission from Suci, P. A., Berglund, D. L., Liepold, L., Brumfield, S., Pitts, B., Davison, W., Oltrogge, L., Hoyt, K. O., Codd, S., Stewart, P. S., Young, M., and Douglas, T. (2007) High-density targeting of a viral multifunctional nanoplatform to a pathogenic, biofilm-forming bacterium, *Chem. Biol.*, **14**(4), 387–398.

Small drug molecules can be covalently attached or encapsulated within VNPs. The chemotherapeutics hygromycin and doxorubicin have been covalently attached to M13. Attachment was at the main coat protein pVIII. Targeting was achieved via presentation of antibodies against ERGR and ErbB2 (both are receptors that are overexpressed on tumor cells); the antibodies were attached to the minor coat protein pIII. Cancer-cell-specific targeting and cell killing were confirmed *in vitro* (Bar *et al.*, 2008).

Besides covalently attaching drugs to the VNP surface, the drug can also be encapsulated into the VNP interior's cavity, thus protecting the molecules. The drug doxorucicin has been encapsulated in RCNMV and HCRSV (recall Chapter 5) (Loo *et al.*, 2008; Ren *et al.*, 2007). Combining targeting and therapy, doxorubicin-encapsulating FA-targeted VNPs have been made using HCRSV. *In vitro* studies confirmed specific cellular uptake and cytoxicity (Ren *et al.*, 2007).

Similarly, dual functionalized MS2 particles were constructed and tested. Here, the drug was encapsulated by covalently linking it to the RNA stem loop operator (recall Section 5.1.1). Encapsulation of the toxin ricin A and the nucleotide analog 5-fluoruoridine was achieved; 90 molecules per particle can be packed (the RNA stem loops bind to every coat protein dimer; MS2 consist of 90 dimers) (Brown *et al.*, 2002). Specific targeting was facilitated using either specific antibodies or Tf. Cell delivery and cytotoxity were shown *in vitro* (Brown *et al.*, 2002).

In summary, many different strategies have been developed that allow the fabrication of dual-functionalized VNPs with potential for site-specific drug delivery. Future studies will undoubtedly provide insights into the performance and efficacy of such materials *in vivo*.

8.6 GENE DELIVERY AND GENE SILENCING

Genetic disorders may involve the over- or underexpression of endogenous genes, and therapy then becomes a matter of silencing or complementing endogenous gene expression. Silencing strategies involve either direct steric inhibition of the translational machinery by the introduction of anti-sense RNA strands complementary to the *messenger RNA* (mRNA), or targeting the mRNA transcript for degradation via the Dicer pathway using *small interfering* RNAs (siRNAs). Upregulation by contrast involves directly introducing genes into the patient for subsequent expression.

Cellular targeting of either strategy has been employed in viral and non-viral systems. Non-viral methodologies include liposomal or polymeric coated nucleic acids, whereas much of the literature among viral targeting methods has focused on the potential of Ads, AAV, and *Lentivirus*[2]. The use of virus-mediated gene delivery is an extensive field; a recent PubMed search yielded thousands of articles. The reader is referred to the following excellent reviews (Ali *et al.*, 1994; Barzon *et al.*, 2005; Benihoud *et al.*, 1999; Choi *et al.*, 2005; Cockrell & Kafri, 2007; Douglas, 2004; Kafri, 2004; Kapturczak *et al.*, 2001; Liu & Kirn, 2008; Sharma *et al.*, 2009; Tenenbaum *et al.*, 2004).

Mammalian vectors such as the aforementioned, however, bear at least the theoretic potential to revert to a pathogenic state. Therefore, employing bacteriophages or plant VNPs in targeting strategies becomes an attractive option. The power of phage-display-mediated targeting to vascular ligands has also been paired with the genome integration functions of AAV to result in particles that are an interesting hybrid between classic gene delivery and nanotechnology approaches. An AAV-targeting vector was packaged within a targeted fd bacteriophage to achieve tissue-specific gene

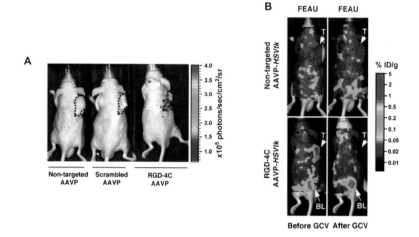

Figure 8.11 Targeted AAV vector-mediated molecular imaging of tumor-bearing mice. (a) *In vivo* bioluminescent imaging (BLI) of luciferase (Luc) expression after systemic AAVP delivery (AAVP denotes a specifically engineered version of AAV, the reader is referred to the original paper, see reference below). Nude mice bearing DU145-derived tumor xenografts received an intravenous single dose of either RGD-4C AAVP-Luc or controls (non-targeted AAVP-Luc or scrambled RGD-4C AAVP-Luc). Ten days later, BLI of tumor-bearing mice was performed. (b) Multitracer PET imaging in tumor-bearing mice after systemic delivery of RGD-4C AAVP-HSVtk. HSVtk = *Herpes simplex virus* thymidine kinase, a reporter gene that can serve for imaging when used in combination with radiol-labeled nucleosides [18F]FEAU or as a suicide gene when used in combination with the prodrug ganciclovir (GCV). Nude mice bearing DU145-derived tumor xenografts (*n* = 9 tumor-bearing mice per cohort) received an intravenous single dose of RGD-4C AAVP-HSVtk or non-targeted AAVP-HSVtk. PET images with [18F]FEAU obtained before and after GCV treatment are presented. T, tumor; BL, bladder. Calibration scales are provided in (a) and (b). Superimposition of PET on photographic images of representative tumor-bearing mice was performed to simplify the interpretation of [18F]FEAU biodistribution. Reproduced with permission from Hajitou, A., Trepel, M., Lilley, C. E., Soghomonyan, S., Alauddin, M. M., Marini, F. C., 3rd, Restel, B. H., Ozawa, M. G., Moya, C. A., Rangel, R., Sun, Y., Zaoui, K., Schmidt, M., von Kalle, C., Weitzman, M. D., Gelovani, J. G., Pasqualini, R., and Arap, W. (2006) A hybrid vector for ligand-directed tumor targeting and molecular imaging, *Cell*, **125**(2), 385–398.

transfer for imaging and therapy (Hajitou *et al.*, 2006b; Trepel *et al.*, 2009). In short, the AAV genome was genetically modified with various reporter gene cassettes (a reporter gene encodes a marker that is easily detectable, such as fluorescent proteins), incorporated into the fd phage genome, and packaged into the bacteriophage. To facilitate targeting of cancer cells and endothelium, the well-established RGD motif was utilized on the bacteriophage exterior. Ligand-directed internalization, and transfer and expression of the reporter genes were demonstrated *in vitro* and *in vivo*. Tissue specificity and efficiency of gene delivery and expression post-systemic administration was confirmed using several tumor mouse models (Hajitou *et al.*, 2006b; Trepel *et al.*, 2009). *In vivo* imaging of reporter genes was performed using bioluminescence imaging and PET (Fig. 8.11). To test the therapeutic potential of the vector, suicide genes were delivered; transgene expression was found to be indeed sufficiently high for effective tumor treatment (Fig. 8.11) (Hajitou *et al.*, 2006b). This study clearly highlights the potential of viral nanotechnology for the development of novel imaging modalities and/or therapies.

In another example, VLPs from the bacteriophage MS2 were engineered to package anti-sense RNA against the 5′-untranslated region and internal ribosome entry site of the *Hepatitis C virus* (HCV). Packaging was accomplished by fusion of the anti-sense RNA sequence to the stem loop triggering encapsulation (recall Section 5.1.1). Cell delivery and penetration was achieved making use of cell-penetrating peptides that were chemically attached to surface Lys side chains present on the VLP. Inhibitory effects on gene expression of HCV were shown *in vitro* using an established reporter system (Wei *et al.*, 2009). The utility of non-human pathogens as viral vectors for gene or RNAi delivery opens a new sector in viral nanotechnology. Future studies evaluating transfection efficiency as well as non-desired side effects comparing non-human viral vectors, viral vectors, and non-viral delivery systems are expected to give further insights into the feasibility of utilizing non-human VLPs for gene delivery.

8.7 FUTURE DIRECTIONS

Recent advances in nanotechnology have led to the development of VNPs and VLPs for potential applications as vaccines, imaging modalities, and targeted therapeutic devices. The vast majority of studies conducted so far, however, are on a biochemical level or in tissue culture *in vitro*. Only a few systems have been evaluated *in vivo*, and information on the *in vivo* performance of specifically engineered and designed VNPs is limited. Studies reported to date demonstrate proof of concept and underline the high potential of VNPs and VLPs as novel candidate materials for medical devices. The next hurdle will be to gain a better understanding of the physiological fate and potential long-term side effects of VNPs. It is anticipated that further development of VNPs for applications in medicine will be an exciting field to follow.

References

Acosta-Ramirez, E., Perez-Flores, R., Majeau`, N., Pastelin-Palacios, R., Gil-Cruz, C., Ramirez-Saldana, M., Manjarrez-Orduno, N., Cervantes-Barragan, L., Santos-Argumedo, L., Flores-Romo, L., Becker, I., Isibasi, A., Leclerc, D., and Lopez-Macias, C. (2008) Translating innate response into long-lasting antibody response by the intrinsic antigen-adjuvant properties of papaya mosaic virus, *Immunology*, **124**(2), 186–197.

Aime, S., Cabella, C., Colombatto, S., Geninatti Crich, S., Gianolio, E., and Maggioni, F. (2002) Insights into the use of paramagnetic Gd(III) complexes in MR-molecular imaging investigations, *J. Magn. Reson. Imaging*, **16**(4), 394–406.

Ali, M., Lemoine, N. R., and Ring, C. J. (1994) The use of DNA viruses as vectors for gene therapy. *Gene Ther.*, **1**(6), 367–384.

Allen, M., Bulte, J. W., Liepold, L., Basu, G., Zywicke, H. A., Frank, J. A., Young, M., and Douglas, T. (2005) Paramagnetic viral nanoparticles as potential high-relaxivity magnetic resonance contrast agents, *Magn. Reson. Med.*, **54**(4), 807–812.

Anderson, E. A., Isaacman, S., Peabody, D. S., Wang, E. Y., Canary, J. W., and Kirshenbaum, K. (2006) Viral nanoparticles donning a paramagnetic coat: conjugation of MRI contrast agents to the MS2 capsid, *Nano Lett.*, **6**(6), 1160–1164.

Arap, W., Haedicke, W., Bernasconi, M., Kain, R., Rajotte, D., Krajewski, S., Ellerby, H. M., Bredesen, d. E., Pasqualini, R., and Ruoslahti, E. (2002) Targeting the prostate for destruction through a vascular address, *Proc. Natl Acad. Sci. USA*, **99**, 1527–1531.

Bar, H., Yacoby, I., and Benhar, I. (2008) Killing cancer cells by targeted drug-carrying phage nanomedicines, *BMC Biotechnol.*, **8**, 37.

Barzon, L., Stefani, A. L., Pacenti, M., and Palu, G. (2005) Versatility of gene therapy vectors through viruses, *Expert Opin. Biol. Ther.*, **5**(5), 639–662.

Bawarski, W. E., Chidlowsky, E., Bharali, D. J., and Mousa, S. A. (2008) Emerging nanopharmaceuticals, *Nanomedicine*, **4**(4), 273–282.

Becker, A., Riefke, B., Ebert, B., Sukowski, U., Rinneberg, H., Semmler, W., and Licha, K. (2000) Macromolecular contrast agents for optical imaging of tumors: comparison of indotricarbocyanine-labeled human serum albumin and transferrin, *Photochem. Photobiol.*, **72**(2), 234–241.

Ben-Gary, H., McKinney, R. L., Rosengart, T., Lesser, M. L., and Crystal, R. G. (2002) Systemic interleukin-6 responses following administration of adenovirus gene transfer vectors to humans by different routes, *Mol. Ther.*, **6**(2), 287–297.

Benihoud, K., Yeh, P., and Perricaudet, M. (1999) Adenovirus vectors for gene delivery, *Curr. Opin. Biotechnol.*, **10**(5), 440–447.

Brabletz, T., Hlubek, F., Spaderna, S., Schmalhofer, O., Hiendlmeyer, E., Jung, A., and Kirchner, T. (2005) Invasion and metastasis in colorectal cancer: epithelial–mesenchymal transition, mesenchymal–epithelial transition, stem cells and beta-catenin, *Cells Tissues Organs*, **179**(1–2), 56–65.

Brennan, F. R., Jones, T. D., Longstaff, M., Chapman, S., Bellaby, T., Smith, H., Xu, F., Hamilton, W. D., and Flock, J. I. (1999) Immunogenicity of peptides derived from a fibronectin-binding protein of *S. aureus* expressed on two different plant viruses, *Vaccine*, **17**(15–16), 1846–1857.

Bridges, K. R., and Smith, B. R. (1985) Discordance between transferrin receptor expression and susceptibility to lysis by natural killer cells, *J. Clin. Invest.*, **76**(3), 913–918.

Brown, W. L., Mastico, R. A., Wu, M., Heal, K. G., Adams, C. J., Murray, J. B., Simpson, J. C., Lord, J. M., Taylor-Robinson, A. W., and Stockley, P. G. (2002) RNA bacteriophage capsid-mediated drug delivery and epitope presentation, *Intervirology*, **45**(4–6), 371–380.

Bruckman, M. A., Kaur, G., Lee, L. A., Xie, F., Sepulveda, J., Breitenkamp, R., Zhang, X., Joralemon, M., Russell, T. P., Emrick, T., and Wang, Q. (2008) Surface modification of tobacco mosaic virus with "click" chemistry, *ChemBioChem*, **9**(4), 519–523.

Cafolla, A., Dragoni, F., Girelli, G., Tosti, M. E., Costante, A., De Luca, A. M., Funaro, D., and Scott, C. S. (2002) Effect of folic acid and vitamin C supplementation on folate status and homocysteine level: a randomised controlled trial in Italian smoker-blood donors, *Atherosclerosis*, **163**(1), 105–111.

Cai, W., and Chen, X. (2007) Nanoplatforms for targeted molecular imaging in living subjects, *Small*, **3**(11), 1840–1854.

Canizares, M. C., Nicholson, L., and Lomonossoff, G. P. (2005) Use of viral vectors for vaccine production in plants, *Immunol. Cell. Biol.*, **83**(3), 263–270.

Chatterji, A., Ochoa, W., Paine, M., Ratna, B. R., Johnson, J. E., and Lin, T. (2004) New addresses on an addressable virus nanoblock: uniquely reactive Lys residues on cowpea mosaic virus, *Chem. Biol.*, **11**(6), 855–863.

Chen, J., Li, R., Yan, S., Li, Q., Bai, T., and Wang, S. (1998) Analysis of the characteristics of folate binding proteins and its relationship with expression of multidrug resistance P-glycoprotein in myelodysplastic syndromes, *Chin. Med. J. (Engl)*, **111**(3), 235–238.

Choi, V. W., McCarty, D. M., and Samulski, R. J. (2005) AAV hybrid serotypes: improved vectors for gene delivery, *Curr. Gene Ther.*, **5**(3), 299–310.

Christ, M., Louis, B., Stoeckel, F., Dieterle, A., Grave, L., Dreyer, D., Kintz, J., Ali Hadji, D., Lusky, M., and Mehtali, M. (2000) Modulation of the inflammatory properties and hepatotoxicity of recombinant adenovirus vectors by the viral E4 gene products, *Hum. Gene Ther.*, **11**(3), 415–427.

Cockrell, A. S., and Kafri, T. (2007) Gene delivery by lentivirus vectors, *Mol. Biotechnol.*, **36**(3), 184–204.

Cornuz, J., Zwahlen, S., Jungi, W. F., Osterwalder, J., Klingler, K., van Melle, G., Bangala, Y., Guessous, I., Muller, P., Willers, J., Maurer, P., Bachmann, M. F., and Cerny, T. (2008) A vaccine against nicotine for smoking cessation: a randomized controlled trial, *PLoS One*, **3**(6), e2547.

Cotter, M. J., and Muruve, D. A. (2005) The induction of inflammation by adenovirus vectors used for gene therapy, *Front. Biosci.*, **10**, 1098–1105.

Dalsgaard, K., Uttenthal, A., Jones, T. D., Xu, F., Merryweather, A., Hamilton, W. D., Langeveld, J. P., Boshuizen, R. S., Kamstrup, S., Lomonossoff, G. P., Porta, C., Vela, C., Casal, J. I., Meloen, R. H., and Rodgers, P. B. (1997) Plant-derived vaccine protects target animals against a viral disease, *Nat. Biotechnol.*, **15**(3), 248–252.

Destito, G., Yeh, R., Rae, C. S., Finn, M. G., and Manchester, M. (2007) Folic acid-mediated targeting of cowpea mosaic virus particles to tumor cells, *Chem. Biol.*, **14**(10), 1152–1162.

Doan, L. X., Li, M., Chen, C., and Yao, Q. (2005) Virus-like particles as HIV-1 vaccines, *Rev. Med. Virol.*, **15**(2), 75–88.

Douglas, J. T. (2004) Adenovirus-mediated gene delivery: an overview, *Methods Mol. Biol.*, **246**, 3–14.

Du, X. L., Zhang, T. L., Yuan, L., Zhao, Y. Y., Li, R. C., Wang, K., Yan, S. C., Zhang, L., Sun, H., and Qian, Z. M. (2002) Complexation of ytterbium to human transferrin and its uptake by K562 cells, *Eur. J. Biochem.*, **269**(24), 6082–6090.

Engler, H., Machemer, T., Philopena, J., Wen, S. F., Quijano, E., Ramachandra, M., Tsai, V., and Ralston, R. (2004) Acute hepatotoxicity of oncolytic adenoviruses in mouse models is associated with expression of wild-type E1a and induction of TNF-alpha, *Virology*, **328**(1), 52–61.

Eto, Y., Yoshioka, Y., Mukai, Y., Okada, N., and Nakagawa, S. (2008) Development of PEGylated adenovirus vector with targeting ligand, *Int. J. Pharm.*, **354**(1–2), 3–8.

Evans, R. M. (1998) Vimentin: the conundrum of the intermediate filament gene family, *Bioessays*, **20**(1), 79–86.

Everts, M., Saini, V., Leddon, J. L., Kok, R. J., Stoff-Khalili, M., Preuss, M. A., Millican, C. L., Perkins, G., Brown, J. M., Bagaria, H., Nikles, D. E., Johnson, D. T., Zharov, V. P., and Curiel, D. T. (2006) Covalently linked Au nanoparticles to a viral vector: potential for combined photothermal and gene cancer therapy, *Nano Lett.*, **6**(4), 587–591.

Garcea, R. L., and Gissmann, L. (2004) Virus-like particles as vaccines and vessels for the delivery of small molecules, *Curr. Opin. Biotechnol.*, **15**(6), 513–517.

Gatto, D., Ruedl, C., Odermatt, B., and Bachmann, M. F. (2004) Rapid response of marginal zone B cells to viral particles, *J. Immunol.*, **173**(7), 4308–4316.

Gilles, C., Polette, M., Mestdagt, M., Nawrocki-Raby, B., Ruggeri, P., Birembaut, P., and Foidart, J. M. (2003) Transactivation of vimentin by beta-catenin in human breast cancer cells, *Cancer Res.*, **63**(10), 2658–2664.

Gomme, P. T., McCann, K. B., and Bertolini, J. (2005) Transferrin: structure, function and potential therapeutic actions, *Drug Discov. Today*, **10**(4), 267–273.

Gonzalez, M. J., Plummer, E. M., Rae, C. S., and Manchester, M. (2009) Interaction of Cowpea mosaic virus (CPMV) nanoparticles with antigen presenting cells in vitro and in vivo, *PLoS One*, **4**(11), e7981.

Graves, E. E., Weissleder, R., and Ntziachristos, V. (2004) Fluorescence molecular imaging of small animal tumor models, *Curr. Mol. Med.*, **4**(4), 419–430.

Grgacic, E. V., and Anderson, D. A. (2006) Virus-like particles: passport to immune recognition, *Methods*, **40**(1), 60–65.

Hajitou, A., Pasqualini, R., and Arap, W. (2006a) Vascular targeting: recent advances and therapeutic perspectives, *Trends Cardiovasc. Med.*, **16**(3), 80–88.

Hajitou, A., Trepel, M., Lilley, C. E., Soghomonyan, S., Alauddin, M. M., Marini, F. C., III, Restel, B. H., Ozawa, M. G., Moya, C. A., Rangel, R., Sun, Y., Zaoui, K., Schmidt, M., von Kalle, C., Weitzman, M. D., Gelovani, J. G., Pasqualini, R., and Arap, W. (2006b) A hybrid vector for ligand-directed tumor targeting and molecular imaging, *Cell*, **125**(2), 385–398.

Hardman, R. (2006) A toxicologic review of quantum dots: toxicity depends on physicochemical and environmental factors, *Environ. Health Perspect.*, **114**(2), 165–172.

Harris, J. M., and Chess, R. B. (2003) Effect of pegylation on pharmaceuticals, *Nat. Rev. Drug Discov.*, **2**(3), 214–221.

Higginbotham, J. N., Seth, P., Blaese, R. M., and Ramsey, W. J. (2002) The release of inflammatory cytokines from human peripheral blood mononuclear cells in vitro following exposure to adenovirus variants and capsid, *Hum. Gene Ther.*, **13**(1), 129–141.

Hooker, J. M., Datta, A., Botta, M., Raymond, K. N., and Francis, M. B. (2007) Magnetic resonance contrast agents from viral capsid shells: a comparison of exterior and interior cargo strategies, *Nano Lett.*, **7**(8), 2207–2210.

Huang, X., Jain, P. K., El-Sayed, I. H., and El-Sayed, M. A. (2008) Plasmonic photothermal therapy (PPTT) using gold nanoparticles, *Lasers Med. Sci.*, **23**(3), 217–228.

Huet, D., Bagot, M., Loyaux, D., Capdevielle, J., Conraux, L., Ferrara, P., Bensussan, A., and Marie-Cardine, A. (2006) SC5 mAb represents a unique tool for the detection of extracellular vimentin as a specific marker of Sezary cells, *J. Immunol.*, **176**(1):, 652–659.

Huttner, N. A., Girod, A., Perabo, L., Edbauer, D., Kleinschmidt, J. A., Buning, H., and Hallek, M. (2003) Genetic modifications of the adeno-associated virus type 2 capsid reduce the affinity and the neutralizing effects of human serum antibodies, *Gene Ther.*, **10**(26), 2139–2147.

Immordino, M. L., Dosio, F., and Cattel, L. (2006) Stealth liposomes: review of the basic science, rationale, and clinical applications, existing and potential, *Int. J. Nanomed.*, **1**(3), 297–315.

Inoue, T., Cavanaugh, P. G., Steck, P. A., Brunner, N. and, Nicolson, G. L. (1993) Differences in transferrin response and numbers of transferrin receptors in rat and human mammary carcinoma lines of different metastatic potentials, *J. Cell Physiol.*, **156**(1), 212–217.

Jain, K. K. (2008) Nanomedicine: application of nanobiotechnology in medical practice, *Med. Princ. Pract.*, **17**(2):,89–101.

Johnson, J., Lin, T., and Lomonossoff, G. (1997) Presentation of heterologous peptides on plant viruses: genetics, structure, and function, *Annu. Rev. Phytopathol.*, **35**, 67–86.

Kafri, T. (2004) Gene delivery by lentivirus vectors an overview, *Methods Mol. Biol.*, **246**, 367–390.

Kaiser, C. R., Flenniken, M. L., Gillitzer, E., Harmsen, A. L., Harmsen, A. G., Jutila, M. A., Douglas, T., and Young, M. J. (2007) Biodistribution studies of protein cage nanoparticles demonstrate broad tissue distribution and rapid clearance in vivo, *Int. J. Nanomed.*, **2**(4), 715–733.

Kapturczak, M. H., Flotte, T., and Atkinson, M. A. (2001) Adeno-associated virus (AAV) as a vehicle for therapeutic gene delivery: improvements in vector design and viral production enhance potential to prolong graft survival in pancreatic islet cell transplantation for the reversal of type 1 diabetes, *Curr. Mol. Med.*, **1**(2), 245–258.

Kobayashi, H., and Brechbiel, M. W. (2005) Nano-sized MRI contrast agents with dendrimer cores, *Adv. Drug Deliv. Rev.*, **57**(15), 2271–2286.

Kokkinos, M. I., Wafai, R., Wong, M. K., Newgreen, D. F., Thompson, E. W., and Waltham, M. (2007) Vimentin and epithelial-mesenchymal transition in human breast cancer — observations in vitro and in vivo, *Cells Tissues Organs*, **185**(1–3), 191–203.

Koo, M., Bendahmane, M., Lettieri, G. A., Paoletti, A. D., Lane, T. E., Fitchen, J. H., Buchmeier, M. J., and Beachy, R. N. (1999) Protective immunity against murine hepatitis virus (MHV) induced by intranasal or subcutaneous administration of hybrids of tobacco mosaic virus that carries an MHV epitope, *Proc. Natl Acad. Sci. USA*, **96**(14), 7774–7779.

Koudelka, K. J., Destito, G., Plummer, E. M., Trauger, S. A., Siuzdak, G., and Manchester, M. (2009) Endothelial targeting of cowpea mosaic virus via surface vimentin, *PLOS Pathog.*, **5**(5), e1000417.

Koudelka, K. J., Rae, C. S., Gonzalez, M. J., and Manchester, M. (2007) Interaction between a 54-kilodalton mammalian cell surface protein and cowpea mosaic virus, *J. Virol.*, **81**(4), 1632–1640.

Kovacs, E. W., Hooker, J. M., Romanini, D. W., Holder, P. G., Berry, K. E., and Francis, M. B. (2007) Dual-surface-modified bacteriophage MS2 as an ideal scaffold for a viral capsid-based drug delivery system, *Bioconjug. Chem.*, **18**(4), 1140–1147.

Lacasse, P., Denis, J., Lapointe, R., Leclerc, D., and Lamarre, A. (2008) Novel plant virus-based vaccine induces protective cytotoxic T-lymphocyte-mediated antiviral immunity through dendritic cell maturation, *J. Virol.*, **82**(2), 785–794.

Langeveld, J. P., Brennan, F. R., Martinez-Torrecuadrada, J. L., Jones, T. D., Boshuizen, R. S., Vela, C., Casal, J. I., Kamstrup, S., Dalsgaard, K., Meloen, R. H., Bendig, M. M., and Hamilton, W. D. (2001) Inactivated recombinant plant virus protects dogs from a lethal challenge with canine parvovirus, *Vaccine*, **19**(27), 3661–3670.

Leamon, C. P., and Low, P. S. (1992) Cytotoxicity of momordin-folate conjugates in cultured human cells, *J. Biol. Chem.*, **267**(35), 24966–24971.

Leamon, C. P., and Low, P. S. (1993) Membrane folate-binding proteins are responsible for folate-protein conjugate endocytosis into cultured cells, *Biochem. J.*, **291**(Pt 3), 855–860.

Lee, G. K., Maheshri, N., Kaspar, B., and Schaffer, D. V. (2005) PEG conjugation moderately protects adeno-associated viral vectors against antibody neutralization, *Biotechnol. Bioeng.*, **92**(1), 24–34.

Levchenko, T. S., Rammohan, R., Lukyanov, A. N., Whiteman, K. R., and Torchilin, V. P. (2002) Liposome clearance in mice: the effect of a separate and combined presence of surface charge and polymer coating, *Int. J. Pharm.*, **240**(1–2), 95–102.

Lewis, J. D., Destito, G., Zijlstra, A., Gonzalez, M. J., Quigley, J. P., Manchester, M., and Stuhlmann, H. (2006) Viral nanoparticles as tools for intravital vascular imaging, *Nat. Med.*, **12**(3), 354–360.

Li, F., Zhang, Z. P., Peng, J., Cui, Z. Q., Pang, D. W., Li, K., Wei, H. P., Zhou, Y. F., Wen, J. K., and Zhang, X. E. (2009) Imaging viral behavior in Mammalian cells with self-assembled capsid-quantum-dot hybrid particles, *Small*, **5**(6), 718–726.

Li, H., and Qian, Z. M. (2002) Transferrin/transferrin receptor-mediated drug delivery, *Med. Res. Rev.*, **22**(3), 225–250.

Li, H., Sun, H., and Qian, Z. M. (2002) The role of the transferrin-transferrin-receptor system in drug delivery and targeting, *Trends Pharmacol. Sci.*, **23**(5), 206–209.

Li, R., Hoess, R. H., Bennett, J. S., and DeGrado, W. F. (2003) Use of phage display to probe the evolution of binding specificity and affinity in integrins, *Protein Eng.*, **16**(1), 65–72.

Li, S. D., and Huang, L. (2008) Pharmacokinetics and biodistribution of nanoparticles, *Mol. Pharm.*, **5**(4), 496–504.

Liepold, L., Anderson, S., Willits, D., Oltrogge, L., Frank, J. A., Douglas, T., and Young, M. (2007) Viral capsids as MRI contrast agents, *Magn. Reson. Med.*, **58**(5), 871–879.

Liu, T. C., and Kirn, D. (2008) Gene therapy progress and prospects cancer: oncolytic viruses, *Gene Ther.*, **15**(12), 877–884.

Loo, L., Guenther, R. H., Lommel, S. A., and Franzen, S. (2008) Infusion of dye molecules into Red clover necrotic mosaic virus, *Chem. Commun. (Camb)*, (1), 88–90.

Lu, Y., and Low, P. S. (2002a) Folate targeting of haptens to cancer cell surfaces mediates immunotherapy of syngeneic murine tumors, *Cancer Immunol. Immunother.*, **51**(3), 153–162.

Lu, Y., and Low, P. S. (2002b) Folate-mediated delivery of macromolecular anticancer therapeutic agents, *Adv. Drug Deliv. Rev.*, **54**(5), 675–693.

Ludwig, C., and Wagner, R. (2007) Virus-like particles-universal molecular toolboxes, *Curr. Opin. Biotechnol.*, **18**(6), 537–545.

Maham, A., Tang, Z., Wu, H., Wang, J., and Lin, Y. (2009) Protein-based nanomedicine platforms for drug delivery, *Small*, **5**(15), 1706–1721.

Majoros, I. J., Williams, C. R., and Baker, J. R., Jr. (2008) Current dendrimer applications in cancer diagnosis and therapy, *Curr. Top. Med. Chem.*, **8**(14), 1165–1179.

Manayani, D. J., Thomas, D., Dryden, K. A., Reddy, V., Siladi, M. E., Marlett, J. M., Rainey, G. J., Pique, M. E., Scobie, H. M., Yeager, M., Young, J. A., Manchester, M., and Schneemann, A. (2007) A viral nanoparticle with dual function as an anthrax antitoxin and vaccine, *PLoS Pathog.*, **3**(10), 1422–1431.

Manchester, M., and Singh, P. (2006) Virus-based nanoparticles (VNPs): platform technologies for diagnostic imaging, *Adv. Drug Deliv. Rev.*, **58**(14), 1505–1522.

Manchester, M., and Steinmetz, N. F. (2008) *Viruses and Nanotechnology*, vol. 327, Springer Verlag, Berlin Heidelberg.

Maurer, P., Jennings, G. T., Willers, J., Rohner, F., Lindman, Y., Roubicek, K., Renner, W. A., Muller, P., and Bachmann, M. F. (2005) A therapeutic vaccine for nicotine dependence: preclinical efficacy, and Phase I safety and immunogenicity, *Eur. J. Immunol.*, **35**(7), 2031–2040.

Meyer, A., Auernheimer, J., Modlinger, A., and Kessler, H. (2006) Targeting RGD recognizing integrins: drug development, biomaterial research, tumor imaging and targeting, *Curr. Pharm. Des.*, **12**(22), 2723–2747.

Mock, M., and Fouet, A. (2001) Anthrax, *Annu. Rev. Microbiol.*, **55**, 647–671.

Mock, M., and Mignot, T. (2003) Anthrax toxins and the host: a story of intimacy, *Cell Microbiol.*, **5**(1), 15–23.

Molenaar, T. J., Michon, I., de Haas, S. A., van Berkel, T. J., Kuiper, J., and Biessen, E. A. (2002) Uptake and processing of modified bacteriophage M13 in mice: implications for phage display, *Virology*, **293**(1), 182–191.

Moore, A. T., Williams, K. E., and Lloyd, J. B. (1977) The effect of chemical treatments of albumin and orosomucoid on rate of clearance from the rat bloodstream and rate of pinocytic capture of rat yolk sac cultured in vitro, *Biochem. J.*, **164**(3), 607–616.

Mor-Vaknin, N., Punturieri, A., Sitwala, K., and Markovitz, D. M. (2003) Vimentin is secreted by activated macrophages, *Nat. Cell Biol.*, **5**, 59–63.

Nanda, A., and St. Croix, B. (2004) Tumor endothelial markers: new targets for cancer therapy, *Curr. Opin. Oncol.*, **16**, 44–49.

Nedellec, P., Vicart, P., Laurent-Winter, C., Martinat, C., Prevost, M. C., and Brahic, M. (1998) Interaction of Theiler's virus with intermediate filaments of infected cells, *J. Virol.*, **72**(12), 9553–9560.

Nicholas, B. L., Brennan, F. R., Martinez-Torrecuadrada, J. L., Casal, J. I., Hamilton, W. D., and Wakelin, D. (2002) Characterization of the immune response to canine parvovirus induced by vaccination with chimaeric plant viruses, *Vaccine*, **20**(21–22), 2727–2734.

Nieminen, M., Henttinen, T., Merinen, M., Marttila-Ichihara, F., Eriksson, J. E., and Jalkanen, S. (2006) Vimentin function in lymphocyte adhesion and transcellular migration, *Nat. Cell Biol.*, **8**(2), 156–162.

Ortel, B., Shea, C. R., and Calzavara-Pinton, P. (2009) Molecular mechanisms of photodynamic therapy, *Front. Biosci.*, **14**, 4157–4172.

Pasqualini, R., Koivunen, E., and Ruoslahti, E. (1997) Alpha v integrins as receptors for tumor targeting by circulating ligands, *Nat. Biotechnol.*, **15**(6), 542–546.

Peiser, L., Mukhopadhyay, S., and Gordon, S. (2002) Scavenger receptors in innate immunity, *Curr. Opin. Immunol.*, **14**(1), 123–128.

Porta, C., and Lomonossoff, G. P. (1998) Scope for using plant viruses to present epitopes from animal pathogens, *Rev. Med. Vir.*, **8**(1), 25–41.

Portney, N. G., and Ozkan, M. (2006) Nano-oncology: drug delivery, imaging, and sensing, *Anal. Bioanal. Chem.*, **384**(3), 620–630.

Prasuhn, D. E., Jr., Singh, P., Strable, E., Brown, S., Manchester, M., and Finn, M. G. (2008) Plasma clearance of bacteriophage Qbeta particles as a function of surface charge, *J. Am. Chem. Soc.*, **130**(4), 1328–1334.

Prasuhn, D. E., Jr., Yeh, R. M., Obenaus, A., Manchester, M., and Finn, M. G. (2007) Viral MRI contrast agents: coordination of Gd by native virions and attachment of Gd complexes by azide-alkyne cycloaddition, *Chem. Commun.*, (12), 1269–1271.

Prato, M., Kostarelos, K., and Bianco, A. (2008) Functionalized carbon nanotubes in drug design and discovery, *Acc. Chem. Res.*, **41**(1), 60–68.

Qian, Z. M., Li, H., Sun, H., and Ho, K. (2002) Targeted drug delivery via the transferrin receptor-mediated endocytosis pathway, *Pharmacol. Rev.*, **54**(4), 561–587.

Rae, C. S., Khor, I. W., Wang, Q., Destito, G., Gonzalez, M. J., Singh, P., Thomas, D. M., Estrada, M. N., Powell, E., Finn, M. G., and Manchester, M. (2005) Systemic trafficking of plant virus nanoparticles in mice via the oral route, *Virology*, **343**(2), 224–235.

Raja, K. S., Wang, Q., Gonzalez, M. J., Manchester, M., Johnson, J. E., and Finn, M. G. (2003) Hybrid virus-polymer materials. 1. Synthesis and properties of PEG-decorated cowpea mosaic virus, *Biomacromolecules*, **3**, 472–476.

Reddy, J. A., and Low, P. S. (1998) Folate-mediated targeting of therapeutic and imaging agents to cancers, *Crit. Rev. Ther. Drug Carrier Syst.*, **15**(6), 587–627.

Reid, T., Warren, R., and Kirn, D. (2002) Intravascular adenoviral agents in cancer patients: lessons from clinical trials, *Cancer Gene Ther.*, **9**(12), 979–986.

Ren, Y., Wong, S. M., and Lim, L. Y. (2007) Folic acid-conjugated protein cages of a plant virus: a novel delivery platform for doxorubicin, *Bioconjug. Chem.*, **18**(3), 836–843.

Roberts, M. J., Bentley, M. D., and Harris, J. M. (2002) Chemistry for peptide and protein PEGylation, *Adv. Drug Deliv. Rev.*, **54**(4), 459–476.

Roser, M., Fischer, D., and Kissel, T. (1998) Surface-modified biodegradable albumin nano- and microspheres. II: effect of surface charges on in vitro phagocytosis and biodistribution in rats, *Eur. J. Pharm. Biopharm.*, **46**(3), 255–263.

Rudin, M., and Weissleder, R. (2003) Molecular imaging in drug discovery and development, *Nat. Rev. Drug Discov.*, **2**(2), 123–131.

Ruoslahti, E. (2002) Specialization of tumor vasculature, *Nat. Rev. Cancer*, **2**, 83–90.

Ryschich, E., Huszty, G., Knaebel, H. P., Hartel, M., Buchler, M. W., and Schmidt, J. (2004) Transferrin receptor is a marker of malignant phenotype in human pancreatic cancer and in neuroendocrine carcinoma of the pancreas, *Eur. J. Cancer*, **40**(9), 1418–1422.

Sandhiya, S., Dkhar, S. A., and Surendiran, A. (2009) Emerging trends of nanomedicine – an overview, *Fundam. Clin. Pharmacol.*, **23**(3), 263–269.

Sato, Y., Yamauchi, N., Takahashi, M., Sasaki, K., Fukaura, J., Neda, H., Fujii, S., Hirayama, M., Itoh, Y., Koshita, Y., Kogawa, K., Kato, J., Sakamaki, S., and Niitsu, Y. (2000) In vivo gene delivery to tumor cells by transferrin–streptavidin-DNA conjugate, *FASEB J.*, **14**(13), 2108–2118.

Sen Gupta, S., Kuzelka, J., Singh, P., Lewis, W. G., Manchester, M., and Finn, M. G. (2005) Accelerated bioorthogonal conjugation: a practical method for the ligation of diverse functional molecules to a polyvalent virus scaffold, *Bioconjug. Chem.*, **16**(6), 1572–1579.

Sharma, A., Li, X., Bangari, D. S., and Mittal, S. K. (2009) Adenovirus receptors and their implications in gene delivery, *Virus Res.*, **143**(2), 184–194.

Shriver, L. P., Koudelka, K. J., and Manchester, M. (2009) Viral nanoparticles associate with regions of inflammation and blood brain barrier disruption during CNS infection, *J. Neuroimmunol.*, **211**(1–2), 66–72.

Singh, P., Destito, G., Schneemann, A., and Manchester, M. (2006a) Canine parvovirus-like particles, a novel nanomaterial for tumor targeting, *J. Nanobiotechnol.*, **4**, 2.

Singh, P., Gonzalez, M. J., and Manchester, M. (2006b) Virsues and their uses in nanotechnology, *Drug Dev. Res.*, **67**, 23–41.

Singh, P., Prasuhn, D., Yeh, R. M., Destito, G., Rae, C. S., Osborn, K., Finn, M. G., and Manchester, M. (2007) Bio-distribution, toxicity and pathology of cowpea mosaic virus nanoparticles in vivo, *J. Control Release*, **120**(1–2), 41–50.

Sipkins, D. A., Cheresh, D. A., Kazemi, M. R., Nevin, L. M., Bednarski, M. D., and Li, K. C. (1998) Detection of tumor angiogenesis in vivo by alphaVbeta3-targeted magnetic resonance imaging, *Nat. Med.*, **4**(5), 623–626.

Smith, M. L., Corbo, T., Bernales, J., Lindbo, J. A., Pogue, G. P., Palmer, K. E., and McCormick, A. A. (2007) Assembly of trans-encapsidated recombinant viral vectors engineered from Tobacco mosaic virus and Semliki Forest virus and their evaluation as immunogens, *Virology*, **358**(2), 321–333.

Smith, M. L., Lindbo, J. A., Dillard-Telm, S., Brosio, P. M., Lasnik, A. B., McCormick, A. A., Nguyen, L. V., and Palmer, K. E. (2006) Modified tobacco mosaic virus particles as scaffolds for display of protein antigens for vaccine applications, *Virology*, **348**(2), 475–488.

Soussan, E., Cassel, S., Blanzat, M., and Rico-Lattes, I. (2009) Drug delivery by soft matter: matrix and vesicular carriers, *Angew. Chem. Int. Ed. Engl.*, **48**(2), 274–288.

Souza, G. R., Christianson, D. R., Staquicini, F. I., Ozawa, M. G., Snyder, E. Y., Sidman, R. L., Miller, J. H., Arap, W., and Pasqualini, R. (2006) Networks of gold nanoparticles and bacteriophage as biological sensors and cell-targeting agents, *Proc. Natl Acad. Sci. USA*, **103**, 1215–1220.

Srivastava, A. S., Kaido, T., and Carrier, E. (2004) Immunological factors that affect the in vivo fate of T7 phage in the mouse, *J. Virol. Methods*, **115**(1), 99–104.

Steinmetz, N. F., and Evans, D. J. (2007) Utilisation of plant viruses in bionanotechnology, *Org. Biomol. Chem.*, **5**(18), 2891–2902.

Steinmetz, N. F., Hong, V., Spoerke, E. D., Lu, P., Breitenkamp, K., Finn, M. G., and Manchester, M. (2009a) Buckyballs meet viral nanoparticles: candidates for biomedicine, *J. Am. Chem. Soc.*, **131**(47), 17093–17095.

Steinmetz, N. F., Lin, T., Lomonossoff, G. P., and Johnson, J. E. (2009b) Structure-based engineering of an icosahedral virus for nanomedicine and nanotechnology, *Curr. Top. Microbiol. Immunol.*, **327**, 23–58.

Steinmetz, N. F., and Manchester, M. (2009) PEGylated viral nanoparticles for biomedicine: the impact of PEG chain length on vnp cell interactions in vitro and ex vivo, *Biomacromolecules*, **10**(4): 784–792.

Steinmetz, N. F., Mertens, M. E., Taurog, R. E., Johnson, J. E., Commandeur, U., Fischer, R., and Manchester, M. (2009c) Potato virus X as a novel platform for potential biomedical applications, *Nano Lett* (accepted).

Suci, P. A., Berglund, D. L., Liepold, L., Brumfield, S., Pitts, B., Davison, W., Oltrogge, L., Hoyt, K. O., Codd, S., Stewart, P. S., Young, M., and Douglas, T. (2007a) High-density targeting of a viral multifunctional nanoplatform to a pathogenic, biofilm-forming bacterium, *Chem. Biol.*, **14**(4), 387–398.

Suci, P. A., Varpness, Z., Gillitzer, E., Douglas, T., and Young, M. (2007b) Targeting and photodynamic killing of a microbial pathogen using protein cage architectures functionalized with a photosensitizer, *Langmuir*, **23**(24), 12280–12286.

Takagi, A., Hirose, A., Nishimura, T., Fukumori, N., Ogata, A., Ohashi, N., Kitajima, S., and Kanno, J. (2008) Induction of mesothelioma in p53+/- mouse by intraperitoneal application of multi-wall carbon nanotube, *J. Toxicol. Sci.*, **33**(1), 105–116.

Tenenbaum, L., Chtarto, A., Lehtonen, E., Velu, T., Brotchi, J., and Levivier, M. (2004) Recombinant AAV-mediated gene delivery to the central nervous system, *J. Gene Med.*, **6**(S1), S212–S222.

Thurber, G. M., Schmidt, M. M., and Wittrup, K. D. (2008) Antibody tumor penetration: Transport opposed by systemic and antigen-mediated clearance, *Adv. Drug Deliv. Rev.*, **60**, 1421–1434.

Trepel, M., Stoneham, C. A., Eleftherohorinou, H., Mazarakis, N. D., Pasqualini, R., Arap, W., and Hajitou, A. (2009) A heterotypic bystander effect for tumor cell killing after adeno-associated virus/phage-mediated, vascular-targeted suicide gene transfer, *Mol. Cancer Ther.*, **8**(8), 2383–2391.

Turek, J. J., Leamon, C. P., and Low, P. S. (1993) Endocytosis of folate-protein conjugates: ultrastructural localization in KB cells, *J. Cell Sci.*, **106**(Pt 1), 423–430.

van Beijnum, J. R., Dings, R. P., van der Linden, E., Zwaans, B. M., Ramaekers, F. C., Mayo, K. H., and Griffioen, A. W. (2006) Gene expression of tumor angiogenesis dissected: specific targeting of colon cancer angiogenic vasculature, *Blood*, **108**(7), 2339–2348.

Vitiello, C. L., Merril, C. R., and Adhya, S. (2005) An amino acid substitution in a capsid protein enhances phage survival in mouse circulatory system more than a 1000-fold, *Virus Res.*, **114**(1–2), 101–103.

Wachsmuth, E. D., and Klingmuller, D. (1978) Reduction of plasma clearance rates and immune response by negative charges: lactic dehydrogenase isoenzymes in normal and LDV-infected mice. *J. Reticuloendothel. Soc.*, **24**(3), 227–241.

Wang, N., and Stamenovic, D. (2002) Mechanics of vimentin intermediate filaments, *J. Muscle Res. Cell Motil.*, **23**(5–6), 535–540.

Wang, S., and Low, P. S. (1998) Folate-mediated targeting of antineoplastic drugs, imaging agents, and nucleic acids to cancer cells, *J. Control Release*, **53**(1–3), 39–48.

Wattendorf, U., and Merkle, H. P. (2008) PEGylation as a tool for the biomedical engineering of surface modified microparticles, *J. Pharm. Sci.*, **97**(11), 4655-4669.

Winter, P. M., Caruthers, S. D., Kassner, A., Harris, T. D., Chinen, L. K., Allen, J. S., Lacy, E. K., Zhang, H., Robertson, J. D., Wickline, S. A., and Lanza, G. M. (2003) Molecular imaging of angiogenesis in nascent Vx-2 rabbit tumors using a novel alpha(nu)beta3-targeted nanoparticle and 1.5 tesla magnetic resonance imaging, *Cancer Res.*, **63**(18), 5838–5843.

Wu, C., Barnhill, H., Liang, X., Wang, Q., and Jiang, H. (2005) A nw probe using hybrid virus-dye nanoparticles for near-infrared fluorescence tomography, *Opt. Commun.*, **255**, 366–374.

Wu, L., Jiang, L., Zhou, Z., Fan, J., Zhang, Q., Zhu, H., Han, Q., and Xu, Z. (2003) Expression of foot-and-mouth disease virus epitopes in tobacco by a tobacco mosaic virus-based vector, *Vaccine*, **21**(27–30), 4390–4398.

Xing, Y., and Rao, J. (2008) Quantum dot bioconjugates for in vitro diagnostics & in vivo imaging, *Cancer Biomark.*, **4**(6), 307–319.

Xu, B., deWaal, R. M., Mor-Vaknin, N., Hibbard, C., Markovitz, D. M., and Kahn, M. L. (2004) The endothelial cell-specific antibody PAL-E identifies a secreted form of vimentin in the blood vasculature, *Mol. Cell Biol.*, **24**(20), 9198–9206.

Yacoby, I., Shamis, M., Bar, H., Shabat, D., and Benhar, I. (2006) Targeting antibacterial agents by using drug-carrying filamentous bacteriophages, *Antimicrob. Agents Chemother.*, **50**(6), 2087–2097.

Yamasaki, Y., Sumimoto, K., Nishikawa, M., Yamashita, F., Yamaoka, K., Hashida, M., and Takakura, Y. (2002) Pharmacokinetic analysis of in vivo disposition of succinylated proteins targeted to liver nonparenchymal cells via scavenger receptors: importance of molecular size and negative charge density for in vivo recognition by receptors, *J. Pharmacol. Exp. Ther.*, **301**(2), 467–477.

Young, M., Willits, D., Uchida, M., and Douglas, T. (2008) Plant viruses as biotemplates for materials and their use in nanotechnology, *Annu. Rev. Phytopathol.*, **46**, 361–384.

Yusibov, V., Rabindran, S., Commandeur, U., Twyman, R. M., and Fischer, R. (2006) The potential of plant virus vectors for vaccine production, *Drugs R. D.*, **7**(4), 203–217.

Chapter 9

SUMMARY AND OUTLOOK

Since the formative studies conducted in the early 1990s, an increasing number of researchers have become interested in the viral nanotechnology field. The interdisciplinary nature of the work, sitting at the interface of virology, chemistry, materials science, and medicine, has facilitated cross-fertilization of ideas and techniques and led to the development of the manifold applications of viral nanoparticles (VNPs) discussed in this book.

A wide variety of chemistries have been developed that facilitated the functionalization and incorporation of VNPs into devices. VNPs can be modified using bioconjugation chemistries, as well as mineralization and metal deposition techniques. VNPs can be arranged into 1D, 2D, and 3D arrays. As a result of these manipulations, VNPs have been fabricated as sensors, tracers, catalysts, memory storage devices, photovoltaic devices, and even battery electrodes. As additional chemistries and VNP platforms become available, it seems that progress in this area is limited only by the imagination.

The principles of self-assembly drive innovation in VLP encapsulation strategies. The self-assembly of VLPs and hybrid systems can be triggered *in vitro* with great flexibility. It is truly remarkable that these viral coat proteins can be assembled around artificial cores, allowing encapsulation of various materials *in vitro*, whereas *in vivo* in the natural host only the viral genome is packaged. Studies to discriminate between material and genome packaging are likely to illuminate the mechanisms underlying both processes. Furthermore, an understanding of the self-assembly mechanisms of VNPs and hybrid VLPs is interesting not only from a materials engineering point of view, but these experimental approaches also provide supporting data to test theoretical hypotheses regarding virus structure and assembly. For example, the assembly of tubular structures using coat proteins that naturally assemble into icosahedrons allows application of the Caspar–Klug theory, described in Chapter 2, to non-icosahedral structures.

Viral Nanoparticles: Tools for Materials Science and Biomedicine
By Nicole F. Steinmetz and Marianne Manchester
Copyright © 2011 by Pan Stanford Publishing Pte. Ltd.
www.panstanford.com

In medicine, VNPs have been applied in diagnostic assays, as vaccines, imaging modalities, and targeted therapeutic devices. This area of research shows great promise, and the next phase of investigation will necessitate a more detailed understanding of the *in vivo* performance of VNPs and hybrid assemblies. The next few years will provide a more sophisticated understanding of the pharmacokinetic and pharmacodynamic properties, toxicity, and efficacy of VNPs in pre-clinical models and clinical trials.

Although a significant number of different viruses have already been exploited as VNPs during the last 20 years, this number is a tiny fraction of the number of viruses available for development. An enormous variety of the viruses that infect plants, bacteria, archaea, and fungi have been structurally and genetically characterized, revealing a vast collection of molecular nanocontainers with different shapes, stabilities, chemical reactivities, and potential for interaction with other nanoplatforms or with cellular or physiologic systems. Furthermore, these characterized viruses in turn represent a small subset of the total number of virus species found in nature. Thus, it is expected that as new viral platforms are evaluated for their utility as VNPs the number of possible uses for VNP technology will expand as well.

Optimization of VNP production and scale-up has already been successful for some systems, such as vaccine production, for example, the *Hepatitis B* and *Human papilloma virus* vaccines. Many other VNPs can be produced on a relatively large scale with ease; however, the chemical functionalization procedures have not yet been scaled up. As the application of VNP technologies progresses toward industrial practice and clinical trials, large-scale production, quality control, reproducibility, and safety become increasingly important.

Viral nanotechnology has grown out of its infancy, and a new era has begun in which pathogens have become tailorable nanoscale building materials. Many studies summarized in this volume emphasize the outstanding potential of VNPs. Viral nanotechnology will continue to be an inspiring and fast-paced field, one that holds great promise for the development of next-generation devices and therapeutics.

Appendix A

PVX and CPMV walk into a bar …

By *Paul Szewczyk*, The Scripps Research Institute, CA, USA.
Drawings by *Yeon-Hee Lim*, Merck Research Laboratories, Merck & Co. Inc., NJ, USA.

Appendix B

What is nano?

Several answers to the question "What is nano?" are given in the movie clip "What has a football to do with nanotechnology?" Watch it using the Web link http://tiny.cc/nanofootball.

Appendix C

Abbreviations

B.1 Viral nanoparticles

Abbreviation	Description
AAV	Adeno-associated virus
ABV	*Acidianus* bottle-shaped virus
Ad	*Adenovirus*
BMV	*Brome mosaic virus*
CarMV	*Carnation mottle virus*
CCMV	*Cowpea chlorotic mottle virus*
CIV	*Chilo iridescent virus*
CPMV	*Cowpea mosaic virus*
CPV	*Canine parvovirus*
FHV	*Flock House virus*
FMDV	*Foot and mouth disease virus*
HBV	*Hepatitis B virus*
HCRSV	*Hibiscus cholorotic ringspot virus*
HCV	*Hepatitis C virus*
HPV	*Human papilloma virus*
PVX	*Potato virus X*
RCNMV	*Red clover necrotic mottle virus*
SIRV2	*Sulfolobus islandicus* rod-shaped virus 2
TMEV	*Theiler's murine encephalomyelitis virus*
TMV	*Tobacco mosaic virus*
TYMV	*Turnip yellow mosaic virus*

B.2 General

Abbreviation	Description
Ab	Antibody
AF	AlexaFluor dye
AFM	Atomic force microscopy
AIDS	Acquired immune deficiency syndrome
ALD	Atomic layer deposition
ATP	Adenosine triphosphate
CAM	Chick chorioallantoic membrane
CAR	Coxsackievirus and adenovirus receptor
cDNA	complementary DNA

CNS	Central nervous system
Cryo-EM	Cryogenic electron microscopy
CuAAC	Copper-catalyzed azide-alkyne cycloaddition chemistry
DIC	Differential interference contrast
DMSO	Dimethyl sulfoxide
DNA	Deoxyribonucleic acid
DPV	Description of plant viruses database
E. coli	Escherichia coli
EDC	1-ethyl-3-(3-dimethylaminipropyl)carbodiimide
EDL	External dielectric layer
EDX	Energy dispersive X-ray spectroscopy
EGF	Epidermal growth factor
ELD	Electroless deposition
FA	Folic acid
FDA	U.S. Government Food & Drug Administration
FGF-2	Fibroblast growth factor
FITC	Fluorescein isothiocynate
FR	Folic acid receptor
FRET	Fluorescence resonance energy transfer
GFP	Green fluorescent protein
GRP	Gastrin-releasing peptide
HAADF STEM	High-angle annular dark-field scanning transmission electron microscopy
HRP	Horseradish peroxidase
HVR	Hypervariable region
IA	Imaging agent
IAA	Iodoacetic acid
ICTV	International Committee on Taxonomy on Viruses
LbL	Layer-by-layer
LED	Light emitting diode
MHA	Mercaptohexanioc acid
MIT	Massachusetts Institute of Technology
MPS	Mercaptopropyl trimethylsilane
MRI	Magnetic resonance imaging
mRNA	Messenger RNA
NCBI	National Center for Biotechnology
NHS	*N*-hydroxysuccinimide
Ni-NTA	Nickel-nitriotriacetic acid
NIR	Near infrared
nm	Nanometer
OAS	Origin of assembly

PAA	Poly(acrylic acid)
PAH	Poly(allylamine hydrochloride)
PAMS	Pathogen-associated molecular patterns
PASA	Polyanetholesulfonic acid
PC	Phosphatidylcholine
PDMS	Polydimethylsiloxane
PDS	Polydextran sulfate
PDT	Photodynamic therapy
PEG	Polyethylene glycol
PEI	Poly(ethyleneimine)
pK_a	Acid dissociation constant
PS	Phosphatidylserine
PSS	Polystyrene sulfonate
PTT	Photothermal therapy
PVA	Polyvinyl alcohol
PVP	Polymer polyvinyl pyrolidone
QD	Quantum dot
RES	Reticuloendothelial system
RNA	Ribonucleic acid
SARS	Severe acute respiratory syndrome
SATP	*N*-succinimidyl-*S*-acetylpropionate
SDS	Sodium dodecylsulfate
SEM	Scanning electron microscopy
Sf	*Spodoptera frugiperda*
SiC	Silicon carbide
siRNA	Small interfering RNA
SPN	Scanning probe nanolithography
StAv	Streptavidin
SVLP-QDs	SV40-like particles with encapsuled quantum dots
SWCNT	Single-walled carbon nanotube
TEM	Transmission electron microscopy
TEOS	Tetraethoxysilane
Tf	Transferrin
TfR	Transferrin receptor
Ti - plasmid	Tumor inducing plasmid
TMR	Tetramethyl rhodamine
TR	Translational repression
UV	Ultraviolet
VIPER	Virus Particle ExploreR
Vir genes	Virulence genes
VLP	Virus-like particle
VNP	Viral nanoparticle

B.3 Amino acids

Amino acid	3-letter code	1-letter code
Alanine	Ala	A
Arginine	Arg	R
Asparagine	Asn	N
Aspartic acid	Asp	D
Cysteine	Cys	C
Glutamic acid	Glu	E
Glutamine	Gln	Q
Glycine	Gly	G
Histidine	His	H
Isoleucine	Ile	I
Leucine	Leu	L
Lysine	Lys	K
Methionine	Met	M
Phenylalanine	Phe	F
Proline	Pro	P
Serine	Ser	S
Threonine	Thr	T
Tryptophan	Try	W
Tyrosine	Tyr	Y
Valine	Val	V

Appendix D

Glossary

Adjuvant	Any substance that can boost or amplify the immune response to an antigen.
Agroinfiltration	A method utilizing *Agrobacterium tumefaciens* to transfer genes of interest into plant genomes for transient expression.
Anisotropic/ Anisotropy	Anisotropy is derived from the Greek word anisos and means "unequal." In materials science, an anisotropic object is directionally dependent. Rod-shaped VNPs have anisotropic properties. Icosahedrons, in contrast, are isotropic; they are homogenous in all directions.
Antibiotics	A chemical compound that kills bacteria.
Antigen	A foreign substance that induces an immune response.
Archaea	Archaea are prokaryotes, which show similarities with bacteria as well as with Eukaryota.
Aspect ratio	The aspect ratio is defined as the length divided by the width of anisotropic objects such as rod-shaped VNPs.
Atomic force microscopy	A high-resolution microscopy technique. It is a physical technique that allows the visualization and manipulation of objects on the nanometer-size scale.
Attenuated virus vaccines	Attenuated viruses are used as vaccines; it is a genetically modified virus strain with reduced virulence.
Bio-orthogonal reactions	The word orthogonal comes from the Greek language and means "straight angle." In chemistry, orthogonal reactions are defined as strategies that allow the deprotection of functional groups independently of each other. Orthogonal reactions are in general highly selective and reactive.
Bioconjugation	The chemical linking of two or more molecules together to form a novel hybrid material.

Biomineralization — A natural process by which living organisms produce minerals.

Biospecific interaction — Biospecific interactions include the binding of an antibody to its target molecule, the interaction between the protein streptavidin and its ligand biotin, or the binding of a ligand to its cell surface receptor.

Biotemplating — The exploration of naturally occurring organic protein structures such as VNPs for the synthesis of man-made materials.

Chemotherapy — The chemical treatment of cancer; the administration of therapeutic drugs.

Chimera — A genetically modified version of the wild-type or the native form of an organism, protein, or VNP. A VNP chimera is, for example, a VNP displaying foreign antigenic peptides. Such chimeras are used in vaccine development.

Chromophores — A chemical structure that is responsible for the coloration of a molecule.

Co-transfection — Transfection of cells that incorporates both a replication-deficient virus and a helper vector or plasmid that supplies the necessary replication proteins in trans.

Coat protein — Proteins that form the exterior "coat" of the virus, which is referred to as the capsid.

Computed tomography (CT) — The word tomography is derived from the Greek word tomos and means "section" or "slice." CT is the computerized processing of tomography images into a 3D image.

Covalent bond — A chemical bond between two molecules in which pairs of electrons are shared. It is the strongest chemical interaction.

Cryo-electron microscopy — A form of electron microscopy where the sample is cooled down to a cryogenic temperature (liquid nitrogen temperature) for analysis. Many virus structures have been studied and solved using cryo-electron microscopy.

Density gradient ultracentrifugation	A separation or purification method, in which molecules are separated based on their density (density = mass per volume).
Diazonium coupling	Diazonium coupling or azo coupling is the reaction between a diazonium salt and a phenol; the phenol could be derived from a tyrosine side chain.
Dip-pen nanolithography	An AFM tip is used as a "pen," which is coated with chemical linkers, the "ink." The tip (or "pen") is brought into contact with the substrate, which can be regarded as the "paper." The molecules (that is the "ink") are transferred onto the surface (the "paper") using solvent meniscus effects. Structures on the nanoscale can be written using this technique.
DNA microarray	DNA microarray technology allows analysis or screening of several thousand DNA molecules on a chip. DNA molecules are first bound on a chip; the immobilized DNA is referred to as target DNA. The DNA chip or DNA array is then probed with RNA or cDNA samples, the probes. If the sequences match, the probe binds the target via base pairing. Detection of hybridization events is typically carried out using fluorescent-labeled probes.
Dynamic light scattering	A spectroscopy technique that utilizes scattered light to determine the size and radius of small particles or proteins.
Electron microscopy	Electron microscopes fire high-energy electrons instead of photons (optical microscopes) toward the sample and create an electronically magnified image that can reach a magnification of $1,000,000 \times$ and allows visualization of objects in the nanometer-size scale.
Electrospinning	Electrospinning uses high voltage to draw fibers from a liquid.
Electrostatic interactions	These are ionic interactions between oppositely charged molecules. Positively and negatively charged molecules are attracted to each other (opposites attract).

Endogenous	An endogenous substance or protein originates from within its host.
Epitope	A small antigenic peptide fragment that is recognized by the immune system.
Expression vector	An expression vector promotes the expression of genes of interest in a non-native host system. Expression vectors can be DNA- or RNA-based. VNP-based expression vectors have also been developed; they are used for the production of pharmaceutically relevant proteins in bacteria or plants.
Fluoro-immunoassay	A fluoro-immunoassay is an antibody-based detection method that takes advantage of the highly specific affinity between an antibody and its antigen. Fluorescent-labeled antibodies or enzyme-labeled antibodies are used to detect the presence of a certain antigen.
Folic acid	A vitamin that is required during growth and development.
Folic acid receptor	Folic acid receptor is the cell surface receptor for the vitamin folic acid; it binds folic acid and thus recruits the vitamin from circulation into the cells where it is needed during growth and development.
Gel electrophoresis	A method utilizing polymerized agarose or acrylamide as a matrix for the separation of charged molecules such as DNA, RNA, or proteins using an electrical field.
Gene therapy	In gene therapy genetic disorders are treated. Gene therapy is the correction of a defective gene. Genetic disorders may involve the over- or under-expression of genes. Therapy then becomes a matter of silencing or complementing endogenous gene expression.
Genotyping	The determination of the genotype (genetic constitution) of an organism.
Heterologous expression system	In heterologous expression, proteins derived from a foreign organism are expressed. For example, plant VNPs can be expressed in bacteria or yeast.
Host-guest encapsulation	A term used in viral nanotechnology to describe the encapsulation of packing of artificial cargo into VNPs. The cargo is the guest and the VNP the host.

Hybrid VNPs	A hybrid VNP is a VNP that is combined with another material, for example, a VNP encapsulating a gold core.
Hydrophilic interactions	Hydrophilic molecules such as water (the word is derived from the Greek language and means "water friendship"). Hydrophilic interactions are mediated via hydrogen bonding; that is, the interaction between hydrogen atoms and electron-negative atoms.
Hydrophobic interaction	Hydrophobicity (the word is derived from the Greek language and means "water fearing") is a physical property; hydrophobic molecules are repelled from water, hydrophobic molecules are attracted to each other (birds of a feather flock together).
Hyperthermophile organism	One that lives in an extremely hot environment. The VNP SIRV2 is an hyperthermophile that inhabits environments with temperatures above 80 °C.
Icosahedron	A polyhedron with 20 triangular faces.
Immunoassay	See "fluoro-immunoassay"
Inactivated virus vaccines	An inactivated virus is a virus that cannot replicate. Inactivation can be carried out by heat or chemical treatment.
Inoculation	The introduction of an infectious organism. Mechanical inoculation is used to propagate plant VNPs. Purified VNPs or infected leaf material is gently rubbed onto leaves of young developing plants.
Light-harvesting systems	Systems that allow the conversion of light into energy.
Magnetic resonance image (MRI)	A medical imaging technique that allows the cross sectioning of human or animal bodies. MRI is based on the alignment of protons from hydrogen atoms in a strong magnetic field. The aligned protons are then exposed to a pulse of radio waves, which leads to absorption of energy. When the second field is turned off, energy is released and can be detected by the scanner to generate the MRI image. Contrast agents are typically used to increase the brightness of the image and hence sensitivity of the technique.

Metamaterials	A synthetic material that gains its functional properties, such as optical and electrical properties, from its structure rather than from its composition.
Mineralization	A technique in which metals or minerals are deposited and nucleated. VNPs have been mineralized with a variety of minerals or metals.
Monodisperse	A monodisperse material is a material that is uniform in shape and size.
Mutant	A genetically modified version of the wild-type organism.
Nanografting	A nanolithographic technique that allows writing structures on the nanometer-size scale. In nanografting an AFM tip is used to remove molecules from a coated surface. First a surface is coated with a chemical (compound A); then the AFM tip is used in contact mode to draw lines or other structures on the surface by removing compound A from the surface. A different functional chemical compound (compound B) can then be introduced onto the surface.
Nanolithography	A technique used to "write" features in the nanometer-size scale on surfaces.
Nanomolding	A nanolithographic technique used to create features such as molds in the nanometer-size scale.
Nematic crystal	A nematic crystal is a liquid crystal. In a nematic crystal molecules have long-range orientational order, but do not show any positional order. The molecules all point in one direction but there is no side-to-side alignment. See also smectic crystal.
Nucleic acid hybridization	The sequence-specific and non-covalent binding between complementary single-stranded nucleic acid strands to form a double-stranded hybrid strand.
Nucleic acid replication	The duplication of a DNA sequence by a DNA polymerase.
Nucleophile	A nucleophile is any atom containing an unshared pair of electrons or an excess of electrons able to participate in covalent bond formations.

Optical imaging	An imaging technology that utilizes photons (light) to create an image.
Orthogonal	See "bio-orthogonal"
Oxime condensation	Oxime condensation describes the chemical reaction between and aldehyde and hydrazide or alkoxyamine.
PEGylation	The attachment of polyethylene glycol (PEG) to biomolecules, VNPs, or other materials.
Phage display technology	A high-throughput screening technique that is used to identify peptides that specifically interact with molecular proteins or certain materials.
Photodynamic therapy	In photodynamic therapy photosensitizers are used that can be activated by radiation. Activated photosensitizers are radicals that can lead to the production of reactive oxygen species, resulting in cell death.
Photothermal therapy	In phototherthermal therapy electromagnetic radiation, such as infrared, is applied to activate a sensitizer in order to release vibrational energy (i.e., heat) to kill target cells.
Photovoltaics	A technology that deals with the conversion of sunlight into electricity.
Physisorption	Physisorption, a.k.a. physical adsorption, is the adsorption of molecules onto surfaces or other molecules caused by van der Waals forces.
Polyvalency	Polyvalency or multivalency describes having a valence of three of more. In chemistry valence is a measure of the number of bonds formed by a molecule. In nanotechnology valency is often referred to as the number of binding sites. An antibody has two binding sites with which to bind target molecules. The antibody is bivalent. A VNP displaying hundreds of targeting ligands could bind to hundreds of target receptors and is thus polyvalent.
Protein cage	A hollow, generally spherical protein structure that is typically assembled by multiple copies of protein monomers.
Quantum dot	A semiconducting crystal of a few nanometers in size.

Reporter gene   A reporter gene encodes a marker that is easily detectable, such as a fluorescent protein.

Reverse transcription   The transcription of single-stranded RNA into the complementary DNA by a reverse transcriptase.

RGD motif   The RGD motif is the peptide containing the amino acid sequence arginine-glycine-asparagine. RGD binds to integrin receptors that are overexpressed on cancer cells and tumor endothelium. RGD is used as a targeting ligand.

Segmented genome   A genome that is divided into several parts. Rather than being encoded in one DNA or RNA strand, the genome is divided into several strands.

Semiconductor   A material that has electrical properties between that of a conductor and an insulator.

Smectic A   See "smectic crystal"

Smectic C   See "smectic crystal"

Smectic crystal   A liquid crystal in which molecules adopt a uniform orientation. The molecules are also ordered in well-defined layers; they are aligned side by side. One further differentiates a smectic A from a smectic C phase. In the smectic A phase the molecules are oriented along the main axis; in a smectic C phase the particles are tilted from the main axis. See also "nematic crystal."

Transcription   Transcription describes the process by which the DNA sequence is translated into a messenger RNA (mRNA) sequence.

Transfection   The introduction of genes into cells using viral vectors or other nanoparticle platforms.

Transferrin   An iron-binding blood plasma glycoprotein, which is responsible for the binding and distribution of iron ions within the body.

Translation   The conversion of the messenger RNA into the amino acid sequence.

Tumor xenografts   A tumor xenograft is generated by transplanting tumor cells of a different species into a host organism; this is typically achieved via subcutaneous injection.

UV/visible spectroscopy	A photometric spectroscopy that uses ultraviolet, visible, and infrared light to determine the fluorescence/absorbance of molecules or proteins.
Van der Waals interactions	These are attractive or repulsive forces between molecules based on dipole effects. Polar chemical compounds have dipole moments. The dipole moment is induced by non-uniform distribution of electron density.
Wet spinning	In wet spinning a solution is extruded through a capillary tube to produce fibers.
X-ray crystallography	A technique used to determine the atomic structure of a molecule or protein complex from a solid crystalline sample.

Appendix E

Acknowledgments

We thank the following colleagues for reviewing chapters for accuracy, and providing helpful comments and discussion:

- *Prof. George P. Lomonossoff*, Professor of Virology, John Innes Centre, Norwich, UK (Chapters 1 and 2).
- *Prof. Anette Schneemann*, Professor of Virology, The Scripps Research Institute, La Jolla, CA, USA (Chapter 3).
- *Mr. Vu Hong*, Graduate Student in Chemistry, The Scripps Research Institute, La Jolla, CA, USA (Chapter 4).
- *Dr. Jon Pokorski*, Postdoctoral Fellow in Chemistry, The Scripps Research Institute, La Jolla, CA, USA (Chapter 4).
- *Prof. Bogdan Dragnea,* Professor of Chemistry, Indiana University, Bloomington, IN, USA (Chapter 5).
- *Prof. Mark Young*, Professor of Virology, Montana State University, Bozeman, MT, USA (Chapter 6).
- *Dr. Dave Evans*, Project Leader in Bionanosciences, John Innes Centre, Norwich, UK (Chapters 6 and 7).
- *Ms. Emily Plummer*, Graduate Student in Viral Nanotechnology, The Scripps Research Institute, La Jolla, CA, USA (Chapter 8).

Special thanks go to the following for assistance in preparing this book:

- *Mr. Paul Szewczyk*, Graduate Student in Structural Biology, The Scripps Research Institute and University of California, San Diego, is thanked for the entertaining joke entitled "PVX and CPMV walk into a bar...". Paul is also thanked for editing several parts of this book.
- *Dr. Yeon-Hee Lim*, Postdoctoral Researcher at Merck Research Laboratories in Merck & Co., Inc., is thanked for the creative cartoon entitled "PVX and CPMV walk into a bar...".
- *Mr. Zhuojun Wu*, Diploma (equivalent to Masters) Student in Biology, RWTH-Aachen University and University of California, San Diego, is thanked for preparation of the abbreviations list and glossary.
- *Mr. Eliskhan Sheripon*, Bachelor Student in Computer Science, University of California, San Diego, is thanked for critical reading and help with the preparation of the glossary.
- *Dr. Vu Hong*, Graduate Student in Chemistry, The Scripps Research Institute, is thanked for providing ChemDraw Figures shown in Chapter 4.

Special thanks also go to our following collaborators and friends:

Emily Plummer, Kristopher Koudelka, Giuseppe Destito, Leah Shriver, Diane Thomas, Mayra Estrada, Pratik Singh, Maria Gonzalez, Chris Rae, Ing Wei Khor, Miriam Berba, John Young, Dave Evans, George Lomonossoff, Elaine Barclay, Kim Findlay, Grant Calder, M.G. Finn, Vu Hong, Chip Breitenkamp, Stanislav Presolski, Petr Cigler, Jon Pokorski, Jolene Lau, Marisa Hovlid, The Finn Lab, Burkhardt Laufer Horst Kessler, Zhuojun Wu, Marianne Mertens, Uli Commandeur, Rainer Fischer, Jack Johnson, Rick Huang, Rebecca Taurog, Jamie Phelps, Eliskhan Sheripon, Olive Ireland, Erik Spoerke, Ralf Richter, Joachim Spatz, Eva Bock, Roger Parker, Tim Noel, Ariane Bize, David Prangishvili, Anette Schneemann, Juan Jovel, Arno Venter, Phil Dawson, Florence Brunel, Anouk Dirksen, Juan Blanco-Canosa, John Lewis, Heidi Stuhlmann, Thomas Ichim, Malene Hansen, Stefan Steinmetz, Lisa Steinmetz, Christian Steinmetz, Franziska Willmes, Joseph Willmes, Yeon-Hee Lim, Paul Szewczyk, Mark Forster, Peter Teriete, Team France, Becky Tennant, Lukas Maurer, Jochen Maurer, Ute Maurer, Nicolas Raabe, Frank Sinatra, A.K., The Spot, The Beach, The Condo, Xanadu, Table Tops, The VNPs, and last but not least CPMV.

The National Institute of Health provides support for our work:

K99EB009105 to N.F.S.

R01CA112075 to M.M.

Index